Re-inventing Drug Development

Edited by

Jeffrey S. Handen, PhD

Medidata Solutions, Inc.
New York, New York, USA

CRC Press
Taylor & Francis Group
Boca Raton London New York

CRC Press is an imprint of the
Taylor & Francis Group, an **informa** business

First published in paperback 2024

First published 2015 by CRC Press
2385 NW Executive Center Drive, Suite 320, Boca Raton FL 33431

and by CRC Press
4 Park Square, Milton Park, Abingdon, Oxon, OX14 4RN

First issued in hardback 2019

CRC Press is an imprint of Taylor & Francis Group, LLC

ISBN: 978-1-4665-7998-9 (hbk)
ISBN: 978-1-03-292999-6 (pbk)
ISBN: 978-0-429-07123-2 (ebk)

DOI: 10.1201/b17594

Visit the Taylor & Francis Web site at
http://www.taylorandfrancis.com

and the CRC Press Web site at
http://www.crcpress.com

Dedicated to my family, Connie, Alex, Max, and Julia Handen

Note

One hundred percent of all authors' royalties from this book are being donated to the National Organization for Rare Disorders, https://www.rarediseases.org/, Washington, D.C., U.S.A. and EURODIS—The Voice of Rare Disease Patients in Europe, http://www.eurordis.org/, Paris, France.

Contents

Editor

Jeffrey S. Handen, PhD, is vice president of professional services with Medidata Solutions, Inc. Dr. Handen has published in multiple peer-reviewed and business journals, presented at numerous industry conferences and scientific meetings as an invited speaker, and served as past editor-in-chief of the *Industrialization of Drug Discovery* compendium. Before joining Medidata Solutions he was with Merck Research Laboratories for 6 years as a director of portfolio and project management helping to plan, manage, and execute Merck's clinical development portfolio. Prior to Merck, Dr. Handen held management consulting positions with Computer Science Corporation and IBM Business Consulting Services (formerly PricewaterhouseCoopers Consulting). With over 20 years' experience in pharmaceutical and biotechnology research and development, process reengineering, and systems and process implementation, Dr. Handen has also held research and management positions with the University of Pennsylvania and the National Institutes of Health. As vice president of professional services for Medidata Solutions, Dr. Handen is responsible for overseeing clinical development business process integration, solution architecting for optimizing clinical trials design execution and development, implementation of data-driven metrics, and developing and implementing operational metrics to improve clinical research processes. He holds a PhD in neurosciences from George Washington University as well as a BS from Duke University.

Contributors

Christopher Bouton, PhD, earned his BA in neuroscience (magna cum laude) from Amherst College in 1996 and his PhD in molecular neurobiology from Johns Hopkins University in 2001. Between 2001 and 2004 Dr. Bouton worked as a computational biologist at LION Bioscience Research Inc. and Aveo Pharmaceuticals, leading the microarray data analysis functions at both companies. In 2004 he accepted the position of head of integrative data mining for Pfizer and led a group of PhD-level scientists conducting research in the areas of computational biology, systems biology, knowledge engineering, software development, machine learning and large-scale omics data analysis. While at Pfizer, Dr. Bouton conceived of and implemented an organization-wide wiki called Pfizerpedia for which he won the prestigious 2007 William E. Upjohn Award in Innovation. In 2008 Dr. Bouton assumed the position of CEO at Entagen (http: //www.entagen.com), a biotechnology company that provides computational research, analysis, and custom software development services for biomedical organizations. In 2013, Entagen was acquired by Thomson Reuters and Dr. Bouton became the general manager of Entagen for Thomson Reuters. Dr. Bouton is an author of more than a dozen scientific papers and book chapters, and his work has been covered in a number of industry news articles.

Daniel R. Denison, PhD, is professor of management and organization at IMD in Lausanne, Switzerland, and the chairman and founding partner of Denison Consulting, LLC. Prior to joining IMD in 1999, Dr. Denison was an associate professor of organizational behavior and human resource management at the University of Michigan Business School, teaching in MBA, PhD, and executive education programs. Dr. Denison has taught and lived in Asia, Europe, Latin America, and the Middle East. He earned his bachelor's degree from Albion College in psychology, sociology, and anthropology, and his PhD in organizational psychology from the University of Michigan. Dr. Denison's research, teaching, and consulting focus on organizational culture and leadership, and the impact that they have on the performance and effectiveness of organizations. His research has shown a strong relationship between organizational culture and business performance metrics such as profitability, growth, customer satisfaction, and innovation. He has consulted with many leading corporations regarding organizational change, leadership development, and the cultural issues associated with mergers and acquisitions, turnarounds, and globalization. His latest book, with IMD colleague R. Hooijberg, *Leading Culture Change in Global Organizations: Aligning Culture and Strategy*, was published in 2012. He has written four other books, including *Corporate Culture and Organizational Effectiveness*, published by John Wiley in 1990. He is also the author of the Denison Organizational Culture Survey and the Denison Leadership Development Surveys, which have been used by over 5,000 organizations. His articles have appeared in leading journals such as the *Academy of Management Journal*, the *Academy of Management Review*, *Organization Science*, the *Administrative Science Quarterly*, and the *Journal of Organizational Behavior*.

Michael S. Katz earned his BS in electrical engineering and computer science and his MBA in finance and management science from Columbia University. Katz's professional career spanned over thirty years in management consulting, recently retiring as a senior partner at Booz, Allen & Hamilton, Inc. He is a 23-year survivor of multiple myeloma and a 6-year survivor of rectal cancer. He has worked as a patient advocate across a broad spectrum of cancers, in research, education, and support. Highlights include chairing the National Cancer Institute (NCI) Director's Consumer Liaison Group, the Cancer Research Advocates Committee at the ECOG-ACRIN Cancer Research Group, the Association of Cancer Online Resources, and serving on the executive board of the International Myeloma Foundation. Katz is also a past member of the NCI's Multiple Myeloma Steering Committee and Patient Advocate Steering Committee, and a peer reviewer for numerous federal government research grant programs. In these roles, Katz has been privileged to be able to actively contribute to improved outcomes for cancer patients, through in-person, on the phone, and online education and support programs, as well as adding the patient voice to the dialog on cancer research. He was actively involved in the development and conduct of the registration trials for two of the immunomodulatory drugs currently being used to treat multiple myeloma. He is acknowledged as the catalyst for the myeloma trial that replaced high-dose dexamethasone with a safer, equally effective lower dose alternative. Katz is the 2014 recipient of the American Society of Clinical Oncology's Partners In Progress Award.

Arun Kejariwal earned his BE in chemical engineering from Manipal Institute of Technology, India, in 1993; ME in chemical engineering from Widener University, Wilmington, Delaware, in 1997; and MBA from Saint

Joseph's University, Philadelphia, Pennsylvania, in 2006. Kejariwal earned his Six Sigma Black Belt in 2009 and Green Belt in 2008. Between 1993 and 1995, Kejariwal worked as a marketing engineer at Incorporated Engineers Limited, India, and designed and installed, in partnership with a British consulting firm, a high-velocity paper drying system, a first in India. From 1997 to 2001, Kejariwal provided process design and validation consultation to biotech, pharmaceutical, food, semiconductor, and cosmetics industries in the United States. In 2001, he accepted the position of research chemical engineer for Merck, and participated in design, installation, and validation of the first biologics multiproduct facility. He led a cross-functional team to manufacture clinical materials for Cancer Mab, and HIV and HPV vaccines. He was identified as the point person for biologics strategic and business planning, and reported to the head of bioprocess research and development. His work on biologics capacity and economic analysis was key to formation of the Merck BioVentures. In 2007, he transitioned to the strategy and portfolio management group at Merck, where he provided scenario and feasibility analysis to support project- and portfolio-level decisions. He participated in design and implementation of multiple end-to-end processes necessary for portfolio and business management. He led a team to develop a model for pipeline forecasting and modeling, and participated in development of an interactive model for portfolio selection and optimization. In December 2010, Kejariwal joined the Portfolio and Decision Analysis (PDA) group at Pfizer Inc. and has led global cross-functional teams in construction and trade-off analysis of strategic alternatives for inflammation and oncology molecules approaching phase III decisions. As a director in PDA, he is responsible for project- and portfolio-level decision support for the Inflammation Therapeutic Area (TA), and portfolio prioritization and optimization for the Global Innovative Business Segment.

Ia Ko, PhD, is a research consultant at Denison Consulting. She is involved in conducting applied research on various topic areas such as organizational culture and innovation for both academic and industry audiences. She holds a PhD in organizational behavior from Claremont Graduate University and a master's in organizational development from Bowling Green State University. Prior to joining Denison, Dr. Ko was involved in applied research, leadership training development, large-scale change, and program evaluation projects for various organizations. Her past and current research interests include organizational culture, engagement, flow, innovation, safety, and women in leadership.

Lindsey Kotrba, PhD, is the president of Denison Consulting. As president, she is focused on the continual growth and development of Denison Consulting as the global leader in providing high-performance culture and leadership solutions to organizations and consultants. Dr. Kotrba has been a part of the Denison organization since 2006 and was the director of research and development before moving into the president role. She is an active researcher on the topics of organizational culture, leadership, and effectiveness with her published research appearing in outlets such as the *Journal of Business and Psychology, Journal of Vocational Behavior, Human Relations*, and *Advances in Global Leadership*. She is also a regular presenter at conferences such as the Society for Industrial/Organizational Psychology, and the Academy of Management. Dr. Kotrba earned her MA and PhD in industrial/organizational psychology from Wayne State University.

Michael A. Martorelli, CFA, is a director at Fairmount Partners. Martorelli came to Fairmount after serving as managing director of research and senior healthcare analyst at Investec Inc. and its predecessor PMG Capital. Prior to PMG Capital, he spent more than ten years at Janney Montgomery Scott as a research analyst covering companies in various segments of the healthcare industry. In addition to participating in Fairmount's healthcare industry investment banking activities, he contributes to the firm's marketing and business development efforts with periodic industry reports and conference presentations. For most of the past decade, he has focused his writing and speaking efforts on pharmaceutical services companies, including those providing tools and technologies used in preclinical and clinical research functions. Martorelli has participated in conferences and similar programs covering various outsourcing topics sponsored by the Drug Information Association (DIA), the Association for Clinical Research Professionals (ACRP), the Tufts Center for the Study of Drug Development (CSDD), and other organizations. In addition to speaking, he has written for publications such as *Dorland's Medical and Healthcare Marketplace Guide, Contract Pharma, Pharmaceutical Executive,* and *Clinical Research and Regulatory Affairs*. Martorelli earned his MBA and his BS in business administration from Drexel University. For the past several years, he has taught finance and investment courses as a member of Drexel's adjunct faculty.

Michele Pontinen has over 25 years' experience as an employee and a management consultant delivering transformation services to the biopharmaceutical industry and government healthcare institutes. She has delivered strategic, cutting-edge solutions, including successfully bridging the data gap between early and clinical development of ethical drugs and biologics thus reducing timelines to approval by regulatory authorities; negotiating and managing information technology and business services contracts, transforming early manufacturing and clinical development business operations; integrating biomedical and clinical data, enabling the strategic use of information technology within the U.S. federal health industry; developing quality system management strategies, based upon the new U.S. Food and Drug Administration (FDA) inspection approach;

and working with clinical and early development information technology vendors to integrate leading edge technical solutions for preclinical and clinical development.

Melinda S. Shockley, PhD, is an accomplished business development professional and entrepreneur. She co-foundd Innolign Biomedical, LLC, a biomedical company developing innovative 3D micro-organ and tissue systems for drug screening and toxicity assessment. She currently holds the position of president of Innolign and is responsible for corporate strategy and management. Between 2001 and 2010, Dr. Shockley was senior director, business development at Medarex, Inc. where she helped to build its pipeline of therapeutic antibody products by establishing strategic collaborations, in-licensing intellectual property and early-stage technologies, and acquiring assets. Prior to the acquisition of Medarex by Bristol-Myers Squibb Company, Dr. Shockley was responsible for certain Medarex out-licensing initiatives with a focus on defining and directing the company's strategy to maximize the value of its antibody drug conjugate technology platform and related therapeutic products. Dr. Shockely has negotiated approximately one hundred contracts to date with a potential total value of several hundred millon dollars, excluding royalties. Prior to joining Medarex, Dr. Shockley was a licensing associate, life sciences at The Johns Hopkins Unviersity. She also held technology transfer positions at the National Institutes of Health/National Heart, Lung, and Blood Institute and the University of Pennsylvania. She conducted research as an NRSA postdoctoral fellow in pharmacology at the multidisciplinary Institute for Medicine and Engineering at the Unversity of Pennsylvania. Dr. Shockely earned her PhD in pharmaceutical chemistry from the University of California, San Franciso and BA in biology from West Virginia University. Dr. Shockley's diverse research background includes G-protein coupled receptor biology as related to neurodegenerative disease, cancer, and atherosclerosis; general mechanotransduction processes in cell biology and disease; and the cellular regulation of cholesterol production.

Jian Wang earned his PhD in bioengineering from the University of Washington in 1996. After a brief postdoctoral post at Carnegie Mellon University, Dr. Wang joined the biotechnology industry—first with Cellomics and then Physiome Sciences and Paradigm Genetics. Dr. Wang joined BioFortis in 2004 as vice president of product development and was later promoted to president and CEO. Through his tenure in the biopharma industry, Dr. Wang developed several commercial life science informatics products with customers in academia, government, and the biopharmaceutical industry. Dr. Wang delights in helping his customers unleash the power of informatics to increase their productivity in scientific, clinical, and translational research.

chapter one

Redefining innovation

Jeffrey S. Handen

Contents

The innovative biopharmaceutical industry is facing unprecedented challenges as it struggles to cope with a host of factors, highlighted by new and ever-increasing economic pressures. By way of one measure as example, the rate of new molecular entity (NME) approvals (both new drug applications [NDAs] and biologics license applications [BLAs]) by the U.S. Food and Drug Administration (FDA) Center for Drug Evaluation and Research (CDER) has essentially remained flat over the past decade, hovering around 23 NME approvals per year, as has the rate of priority NDA approvals, averaging just around 21 priority approvals per year, since 2004. Paired with patent expiries, the results are declining compound annual growth rates (CAGRs) in the industry, as measured by the 14 large-cap pharmaceutical companies. Industry-wide sales CAGRs declined from 10% in the 1999 to 2004 period to 6.7% in the 2005 to 2009 period to projections of 1.2% through 2014 (Goodman, 2009).

The industry is facing increasing pricing pressures as a result of a number of policy and societal stressors that have unfortunately served to frame the economic discussions of biopharmaceuticals and biopharmaceutical development in terms of cost rather than value. Although total expenditures on pharmaceuticals per capita in the United States have exponentially increased over the past 20 years (OECD, 2011), outcomes and evidence-based medicine approaches have lagged in their development and are still systematically failing to keep track to demonstrate the value of increased healthcare spending on medicines (e.g., reduced hospitalizations, increased quality of life measures, and reduced time missed from work). Additionally, these factors have been exacerbated by the very success of the innovative biopharmaceutical industry itself as more and more innovator drugs that have treated an exceptional diversity of the global disease burden come off patent protection and generic versions are brought to market. For instance, even though world-branded pharmaceutical sales in 2010 were estimated at $286 billion (PhRMA, 2010), the growth of generic drug sales is 4 times higher than overall growth in innovator sales, and represented $107.5 billion in sales in 2010, compared to $73.5 billion in 2006. In 2010, 75.4% of prescriptions written in the United States were for generic drugs, predicted to approach 80% in 2012 (IMS Institute for Healthcare Informatics, 2012). An ever-increasing percentage of the global disease burden is now more than adequately treated by generic drugs. Though there obviously still exists large segments of unmet medical needs and opportunities for significant improvement in existing pharmacological standards of care, the innovator biopharmaceutical industry has not been able to maintain the historic rates of breakthrough innovation.

Innovator companies are increasingly finding themselves left behind in justifying their value proposition in the public's mind. Nowhere is this perhaps more evident than in the increasing use of compulsory licensing of innovative pharmaceuticals, first instituted through the World Trade Organization's (WTO) Trade-Related Aspects of Intellectual Property Rights (TRIPS) Agreement, to produce generic copies of innovative drugs not just for domestic consumption in "national … or other circumstances of extreme urgency" but for export. The goal behind TRIPS—to facilitate life-saving innovation reaching severely economically depressed countries or least-developed countries (LDCs)—though noble in its intent, has been questionable in its execution and outcome, as a recent analysis has shown that approximately 50% of the compulsory licensing episodes between 1995 and 2011 occurred not in LDCs but in upper-middle-income countries (Beall and Kuhn, 2012). This observation is not intended to serve as a discussion or critique of compulsory licensing and the TRIPS Agreement, but rather to suggest that the observed trend of increasing usage of compulsory licensing is bound up in the industry's failure to adequately justify its value proposition.

Increasing resource requirements coupled with rising rates of attrition in the product development lifecycle are also plaguing the industry despite everyone's best efforts to bring efficiencies to the value chain in the guises of technological innovations, process-reengineering efforts, new collaborative partnering, new sourcing strategies, and cultural shifts. The industry average fully capitalized cost (including the cost of attrition) to develop one successful new drug has doubled over the past decade and the clinical development time to develop that drug has continued to increase across that same time period, increasing by approximately 50% (Tufts Center for the Study of Drug Development, 2010). Industry average probability of success (POS) rates from a variety of studies have continued to show decreases. Major company POS from phase I to market has decreased from 10% in 2002–2004 to 5% in 2006–2008 (Arrowsmith, 2012). Between-phase attrition rates as measured from the Pharmaceutical Industry Database (PhID), containing information on research and development (R&D) projects for more than 28,000 compounds investigated since 1990, show phase I attrition rates increasing from approximately 30% to 50% from 1990 to 2004, phase II attrition rates increasing from approximately 45% to 70%, and phase III attrition rates increasing from approximately 20% to 50% over the same time frame (Pammolli et al., 2011). And 50% of NMEs fail to get approval when first submitted to the FDA, with 30% never getting approval (Sacks, 2012).

As discussed earlier, innovation itself has also decreased. This can be measured directly from the trend in decreasing NME approvals by the FDA, despite isolated single-year upticks (FDA, "Summary of NDA Approvals") Additionally the number of NMEs overstates the "innovation" seen in the industry. Though the NME classification does include many new chemical entities (NCEs), defined by the FDA as "a drug that contains no active moiety that has been approved by FDA under section 505(b)," it also includes active moieties closely related to already approved drugs, biological products submitted to CDER regardless of whether the agency previously has approved a related active moiety in a different product, other combination products, and various other administrative classifications (see FDA, "Classification of NME"). Thus NCE, and not NME or NDA, would be a better, truer measure of "innovative" drugs.

Competition, not just from patent expiries but also in the form of shrinking periods of market exclusivity, is also contributing to the woes of the industry. Thirty years ago the mean time from first-in-class U.S. approval to first follow on was 5.1 years. Ten years ago that period of effective marketing exclusivity had dropped to 1.8 years (DiMasi and Paquette, 2004). Also, although the number of drug projects in development continues to rapidly increase, surpassing 10,000 industry-wide by some estimates, up from approximately 6,000 just a decade ago (Pharma R&D Annual Review 2010, Pharmprojects), a cursory examination of the pipelines of the largest innovative biopharma companies shows significant overlap, not surprising as the increasing inefficiencies and subsequent decreasing return on investment (ROI) on R&D is driving portfolio managers to concentrate on a relatively few commercially attractive potential market segments. By way of example, 8 out of 10 of the largest biopharma companies have active development programs for type 2 diabetes and breast cancer; 6 out of 10 have active development programs for rheumatoid and psoriatic arthritis, and acute coronary syndrome; and 5 out of 10 for hepatitis C virus (HCV), chronic obstructive pulmonary disease (COPD), Alzheimer's and non-small cell lung carcinoma (NSCLC). In other words everyone is to a large extent working on the same things. If this trend continues, competitive differentiation will need to increasingly come from being best in class, or sales and marketing, or supply chain considerations and not first in class.

Productivity as measured by output for the innovative biopharmaceutical industry is also a point of discussion. Though there are many possible measures and interpretation of productivity along with multiple confounding factors from the macroenvironment, it must be pointed out that only 11 of 42 members of PhRMA existing in 1988 currently exist as independent entities today (Arrowsmith, 2012). Such has been the state of mergers and acquisitions among the innovators in order to maintain productivity and returns. As a matter of fact the pharmaceutical industry as a whole has delivered declining rates of CAGR, as noted earlier, and been consistently outperformed by other "hi-tech" segments over the past 10 years. For example, the total shareholder return (TSR) over the past 10 years as represented by the S&P 1500 Composite Pharmaceuticals Total Industry Return and S&P 1500 Composite Software Total Industry Return show a return of 78.1% for Software versus 24.3% for Pharma (Handen, 2012).

Finally no discussion of the challenges facing the industry would be complete without an examination of regulatory pressures, both real and perceived. Though regulatory requirements and the regulatory burden placed on drug developers have undoubtedly increased, blaming the current state of clinical development on the regulators is a disservice. While clinical development times over the past 30 years have on average tripled, from 2.5 to 7.5 years, regulatory approval phases have on average remained relatively constant around 1 year (FDA, "CDER User Fee Performance," 2012). Though the regulators are demanding more information, this is the natural evolution of advances brought about through successful medical and clinical research. Where this has been exacerbated is due to the failure of both the industry and regulators to frame the discussion of therapeutics development in a true risk–benefit paradigm, just as they have failed to capture value. There is no such thing as a 100% safe or effective drug. Safety and efficacy are measured in the context of the individual patient's and societal benefits outweighing the risks. Solutions do exist for easing the regulatory burden, while still guaranteeing efficient and effective therapeutics development and commercial return on investment. One such prime example is the 1986 National Childhood Vaccine Injury Act (NCVIA) in the United States, which established a no-fault system for litigating vaccine injury claims, and the National Vaccine Injury Compensation Program (NVICP), which in effect balanced the individual's risk against the societal benefits by creating a mechanism to ensure vaccines could still be brought to market in a commercially viable model without unduly restrictive regulatory requirements on the developers.

All of these forces are conspiring to increase pressures on the commoditization of drug development by eating away at the revenues and margins that have funded original, innovative research and provided returns for both patients and investors, thus eroding industry support for truly novel, often high-risk development. The industry has responded with massive cost-cutting measures characterized by significantly increased outsourcing, offshoring, adoption of point technologies, and external collaboration. And unfortunately, it is still not working to the extent the industry had hoped and patients and society expect. A proof point: From 1996 to 2004 industry average fifth-year sales per $1 billion R&D spent yielded a return of $275 million. By the 2005–2010 time frame that same investment only yielded on average $75 million—a 73% decline. Meanwhile global R&D spent per year went from $65 billion to $125 billion—more than double (Dubin, 2012). Spending more money, as an industry, to put out fewer valuable products is not a sustainable business model. Yet innovative drug development is not a luxury. The winners, and there will be winners, will be those organizations that are able to achieve the operational efficiencies and innovations needed to support scientific innovation. This goal of this book is to do just that: begin a dialogue, to share best practices and perspectives from both within and outside the biopharmaceutical industry, and to serve to redefine innovation so as to focus efforts on bringing better medicines to the patients who need them, faster and more cost effectively.

We must start by redefining the traditional notion of innovation. "Innovation" in the pharmaceutical R&D lexicon, has historically been applied to major advances in therapy and unmet medical needs, which can command premium pricing. However, this fails to take into account other forms of innovation that involve R&D and can be just as effective, paradigm shifting, and profitable.

Financial innovation, beyond just traditional generics, provides a huge opportunity to radically improve standards of care and quality of life. Whether debating the merits of establishing biosimilars development pathways within traditional innovative biopharma companies or resourcing investment into alternate, more efficient pathways of active pharmaceutical ingredient (API) syntheses, decreasing the potential out-of-pockets costs of therapeutics to a patient by perhaps up to 90% can be as innovative to that patient's disease burden as meeting an underserved medical need and as innovative to society as meeting an unmet medical need. One example is provided by the work jointly done by Merck & Co., Inc. and Codexis, Inc., in developing a novel biocatalytic method for sitagliptin. The outcome of this improved synthesis method was a 10% to 13% increase in product yield, decreased raw material requirements, but perhaps even more significantly was a substantial (~19%) decrease in the generation of waste byproducts, all at the industrial scale (Savile et al., 2010). This innovative R&D has a financial return potential equivalent to that of many molecules themselves, thus portfolio managers must take into account total cost of ownership and understand the full value potential when deciding which R&D projects to resource and redefine the traditional notion of an "R&D" project as being defined as a molecule.

Market innovation, such as the use of branded or authorized generics can also be used to extend the lifecycle of innovative drugs. Here the biopharmaceutical industry can borrow best practices and learnings from the consumer products industry where brand awareness and loyalty are routinely leveraged to successfully command premium pricing. While this is just starting to be appreciated by the industry, there is still a long way to go. Patient loyalty programs are growing in popularity in Brazil, Mexico, and other emerging markets. With a few noted exceptions, just about every one of the traditional "Big Pharma" innovators does now have a generic subsidiary, yet for the most part they fail to adopt tried-and-true consumer marketing practices opting to abandon potentially lucrative, albeit generally orders of magnitude less, revenue-generating opportunities. One only has to look at Starbucks for an example of a business model that has successfully turned a commodity into a high margin luxury item that is in fact viewed by many not as a luxury item at all but as a daily requirement of living. And when comparing the average price of a large cup of coffee at around ~$4 daily to say a typical price for a generic statin, available at ~$4 for a 30-day supply at several leading U.S. national retail chains and pharmacies, one cannot help to surmise which industry has had more success in communicating its value proposition, real or perceived. Expanding into "new" markets represents an opportunity for the innovative biopharmaceutical industry to continue to recoup some of its investments in original R&D.

Market innovation must also be defined as shifting the R&D infrastructure to support the revenue growth anticipated outside of the traditional American, Western European, and Japanese "major markets." Although starting to be recognized by the industry's sales and marketing efforts, R&D has yet to be truly globalized outside of these traditional major markets. Many so-called strategic partnerships and some limited R&D centers have been established by most large pharmaceutical companies in certain nontraditional geographies (e.g., China), yet for the most part these remain tactical, nonintegrated forays whose portfolios are often managed separately. By 2015 emerging markets are forecast to account for 30% of the global pharmaceuticals revenues, up from 19% in 2010 (Booz & Co., 2011). R&D efforts need to align with supporting clinical R&D, postmarketed R&D, and pharmacoeconomic research in an integrated way in these geographies and not just as a source of low-cost offshoring opportunities.

Pharmacoeconomic innovation, for example, health economics, is also a significant new force shaping where, how, and with whom R&D budgets are spent. The reality of pharmacoeconomics today is not about just getting registered, it's about getting on the formulary, and in certain regulatory jurisdictions it has become statute (e.g., the National Institute for Health and Care Excellence [NICE] in England and Wales). And pharmacoeconomic R&D must be combined with evidence-based medicine and incorporated into clinical development programs so as to demonstrate not just improved meaningful clinical outcomes for patients significantly different than current (or anticipated!) standards of care, but improved meaningful clinical outcomes whose value can be quantitatively captured. In the United Sates the FDA has long used risk–benefit analyses in regulatory approval decisions. Just recently it released a draft guidance document that outlined a proposed framework for benefit–risk determinations for both the premarket approval (PMA) process for high-risk medical devices and the de novo process for low- and moderate-risk medical devices (FDA, "Factors to Consider," 2012). Though the FDA, unlike NICE, has neither the statute authority nor mission to extend the risk–benefit framework to a cost–benefit framework, these guidance documents demonstrate existing methodologies and approaches that could serve as a precedent for industry discussions in the clinical development phases. Again here is an opportunity for portfolio managers and study designers to take into account the true "total cost of ownership" and understand the full value potential when deciding which R&D projects to resource and redefine the traditional notion of an "R&D" project as being defined as a molecule.

Operational innovation also must be brought to the forefront. A victim of its own success from a history of realizing huge revenue returns and enjoying double-digit growth from the blockbuster model of drug development, the innovative pharmaceutical industry has been able to ignore for too long, at its own peril, the advantages of operational innovation and efficiencies. Concepts such as Lean, Six Sigma, Business Process Re-engineering, Process Control Systems, Performance Management, Balance Scorecards, Strategic Partnering, Supplier Rationalization, and so on, though now for the most part firmly established within innovative pharmaceutical industry R&D departments, were first introduced and well entrenched outside of R&D departments and outside of the industry years, and in some cases, decades earlier. Opportunity costs were often underappreciated with decision makers focusing on bringing an innovative molecule to market rather than bringing an innovative molecule to market in the most innovative manner possible.

Part of realizing the benefits of operational innovation is also realizing the benefits of technical innovation. As discussed earlier in this chapter, despite everyone's best efforts, the product development lifecycle continues to increase in time and cost and risk, and decrease in efficiency. Up to 30% of data collected by industry sponsors are never used in a regulatory submission (Tufts Center for Study of Drug Development, 2009) yet the average CRF length has increased by almost 300% over the past 10 years. Additionally, nearly 40 percent of protocol amendments occur before even the first subject, first dose. And rising complexity in protocol design have helped increase the execution burden on sites by 11% per year (Getz et al., 2011) and negatively affected patient recruitment and retention rates as procedures have become more numerous, more frequent, more complex, and more invasive. Optimizing trial resourcing and execution entails ensuring alignment of all protocols to defined study objectives and likewise eliminating study activities that do not support defined endpoints. This also necessitates benchmarking protocol design to industry procedure costs, usage, and frequency to reduce unneeded complexity and facilitate better resource planning. "Automating" and "enabling" discrete process steps by implementing point solutions has not yielded the overall expected efficiencies and synergies in the drug development process, but merely resulted for the most part in the shifting of bottlenecks. A holistic, integrated approach is required that addresses the entire value chain of the drug development process, explicitly linking the desired medical and commercial outcomes through data collection efforts to project planning and design. This will ensure that all aspects are aligned from day one, managed, and remain in alignment, rather than perpetuating cultural, process, and technology silos that merely automate the already existing inefficiencies thus realizing the full value proposition of technology or "eClinical".

Thus, operational and technical innovation combine to form true value-driven clinical design and support innovative drug development. Indeed innovative drug development is not a luxury. The winners will be those organizations that are able to achieve the operational and technical efficiencies and innovations needed to support scientific innovation.

Innovation must also be redefined in terms of cultural innovation. Although the oft-quoted maxim from Peter Drucker "culture eats strategy for lunch," couldn't be truer; other commentators have added to this sentiment paraphrasing that "culture eats process for lunch too." Culture is perhaps the hardest element within any organization to change and arguably the most important. Cultural innovation, risk taking, change management, and so forth are perhaps the least understood and least appreciated strategic initiatives often undertaken by any organization, in large part due to the challenge of tying cultural change to outcomes and general lack of quantitative robust methodologies in assessing and implementing cultural innovation. For instance, although the innovative biopharmaceutical industry has of late championed risk taking as a way to foster innovation, risk taking alone is really not what is needed or what most managers would pursue on their own. Risk taking must be pursued responsibly in the context of a balanced portfolio where high-risk–high-reward development projects (e.g., unmet medical needs, relatively poorly characterized biological targets, and relatively poorly predictive animal models of disease) are balanced with low-risk–low-reward projects (e.g., lifecycle management, generics, and well-established mechanisms of actions).

"What gets measured gets attention": metrics that matter. Implementing the appropriate metrics within a performance management context can also contribute to building a culture of true, value-added innovation. Metrics should be meaningful, few in number (<20), not create or impose an unreasonable burden of collection, be transparent to the entire organization, and come with consequences. A typical large biopharma company tracks hundreds if not thousands of metrics for multiple departments as well as the corporate level. The litmus test for this effort is twofold: (1) How many of these metrics are ever acted upon. Do they guide actions? (2) How many of these metrics are directly tied to quantifiable value creation? Excepting safety and compliance metrics do they tie into the financial models? One example is discovery output, which is typically measured in terms of the number of preclinical candidates, yet in most companies molecules are not valued until they enter the clinical pipeline or sometimes even the late-stage clinical pipeline. This is a disconnect in driving creation of nonvalued assets. Perhaps a better approach would be to measure and accordingly incentivize discovery managers not on their output but on the outcomes (e.g., successful first in human [FIH] or even successful clinical proof of concept [PoC]). Although it is true that the very nature of science and biological systems limits the degree of predictivity and the extended timelines of drug development would delay financial incentives, the very nature of business demands optimizing resource consumption with value-creating or value-adding activities. Compensation systems can be implemented that take into account the time value of money and the delays between discovery output and clinical value creation. This serves to create a culture where all parties in the entire organization are working toward demonstrable value to the shareholders and customers, that is, patients.

Organizational design is also a lever that can be used to effect cultural innovation. As discussed earlier in the chapter the extant operational siloes that have led to implementation of point solutions at the expense of focusing on the entire value chain are also exacerbated by scientific siloes. Therapeutic area (TA) specialization and organizational design in R&D is common, often resulting in "re-inventing the wheel" where mechanism of actions are often not routinely explored for potential leveraging in other TAs. Drug repositioning efforts are often an afterthought where companies are organized along disease departments and not pathways, and fail to effectively take advantage of more systems biology approaches.

Following on this is a reexamination today's innovative biopharmaceutical companies must make to evaluate their missions and core competencies. Most of today's industry leaders grew over the better half of the past century out of their commitment to improve the lives of patients—their mission statements. Help the patients and the profits will follow. Serving patients was once almost exclusively defined as designing and developing better chemically, and more recently biologically, packaged therapeutic interventions. This author would argue that the mission has not changed, but perhaps the core competencies have failed to keep up with the mission statement. The competitive landscape is littered with companies that thought they could equate "science" with innovations in patient and healthcare chemistry with therapeutics development.

Tomorrow's industry leaders must morph from drug development into healthcare companies. This does not mean abandoning biopharmaceutical research and development but rather adapting and adding to it and contributing to the life sciences ecosystem of the R&D industry, payors, providers, and patients vis-à-vis collaboration, communication, and integration. Value must be assessed, and compensated, not by the commoditized number of pills pushed but rather the value of the outcome to the patient and society. This sea of change will require not just different ways of working and reinterpretation of value by the innovative biopharmaceutical industry but also by payors/reimbursers, physicians and medical community, governments, and patients themselves. And this will require the innovative biopharmaceutical industry to reexamine and link or change where its efforts (and profits) should and can be maximized, for example, deciding for which disease states drugs are the optimal therapeutic intervention and for which they are not; and even for those deciding among innovative research versus perhaps compliance management, standardization, and incorporation of electronic medical records (EMRs)/electronic health records (EHRs) into clinical development for operational efficiencies and Big Data mining, leveraging best-in-class global logistics and supply chain capabilities. These are just a few of the issues tomorrow's leaders must grapple with in order to decide on what will be core capabilities and differentiators of tomorrow's innovative biopharmaceutical industry.

In the chapters that follow contributors drawn from the executive ranks of clinical development practitioners and stakeholders; from biopharmaceutical companies, clinical research organizations, academia, the financial community, and the patient perspective have all come together to provide their expertise and visions with the goal of getting a dialogue going in order to radically improve therapeutics development to get more and better medicines to the patients who need them, as fast as possible in as cost-efficient manner as possible.

References

Arrowsmith, John. "A Decade of Change." *Nature Reviews Drug Discovery* 11 (January 2012): 17–18.
Beall, Read, and Kuhn, Randall. "Trends in Compulsory Licensing of Pharmaceuticals Since the Doha Declaration: A Database Analysis." *PLoS Med* 9, no. 1 (2012): e1001154. DOI:10.1371/journal.pmed.1001154.

Booz & Co. "2012 Healthcare Industry Perspective," 2011.

DiMasi, Joseph, and Paquette, Cherie. "The Economics of Follow-on Drug Research and Development." *Pharmacoeconomics* 22 Suppl. no. l2 (2004): 1–14.

Dubin, Cindy. "Pharma R&D Productivity Drops 70%." *Life Sciences Leader*, 2012.

Getz, Kenneth, Campo, Rafael, and Kaitin, Kenneth. "Variability in Protocol Design Complexity by Phase and Therapeutic Area." *Drug Information Journal* 45, no. 4 (2011): 413–420.

Goodman, Michael. "Pharmaceutical Industry Financial Performance." *Nature Reviews Drug Discovery* 8 (2009): 927–928.

Handen, Jeffrey. "Are Clinical Trials Agile Enough? What Pharma Can Learn from Software Development." *Journal for Clinical Studies* 4, no. 6 (2012).

IMS Institute for Healthcare Informatics. *The Use of Medicines in the United States: Review of 2011.* April 2012.

OECD. *Health at a Glance 2011: OECD Indicators.* OECD Publishing, 2011.

Pammolli, Fabio, Magazzini, Laura, and Riccaboni, Massimo. "The Productivity Crisis in Pharmaceutical R&D." *Nature Reviews Drug Discovery* 10 (2011): 428–438.

Pharmaceutical Research and Manufacturers of America. *Pharmaceutical Industry Profile 2010.* Washington, DC: PhRMA, April 2010.

Sacks, Leonard. "FDA Clinical Investigator Training Course," November 2012.

Savile, Christopher, Janey, Jacob, Mundorff, Emily, et al. "Biocatalytic Asymmetric Synthesis of Chiral Amines from Ketones Applied to Sitagliptin Manufacture." *Science* 329, no. 5989 (2010): 305–309.

Tufts Center for Study of Drug Development. "Trends and Implications of Increasingly Complex Protocol Designs," 2009.

Tufts Center for the Study of Drug Development. "Rising Protocol Complexity, Execution Burden Varies Widely by Phase and TA." *Tufts CSDD Impact Report* 12, no. 3, May/June 2010.

U.S. Food and Drug Administration (FDA). "CDER User Fee Performance and New Drug Approvals, 2011 Summary," January 2012.

U.S. Food and Drug Administration (FDA). "Factors to Consider When Making Benefit-Risk Determinations in Medical Device Premarket Approval and *De Novo* Classifications," March 28, 2012.

U.S. Food and Drug Administration (FDA). "Classification of NME." http://www.fda.gov/drugs/developmentapprovalprocess/druginnovation/default.htm.

U.S. Food and Drug Administration (FDA). "Summary of NDA Approvals & Receipts, 1938 to the Present." http://www.fda.gov/AboutFDA/WhatWeDo/History/ProductRegulation/SummaryofNDAApprovalsReceipts1938tothepresent/default.htm.

chapter two

Collaborations of the future

Melinda S. Shockley

Contents

Innovation through partnerships

We have witnessed extraordinary breakthroughs in medicine over the past century; however, the battles to ward off diseases, disorders, and the ailments that come with aging are not yet over. Stunning successes within the industry have led to the public's high expectations for new and even better and safer medicines at low cost. The pharmaceutical industry is tasked with developing innovative medicines while addressing the pressures of rising costs, increased stringency in drug development processes and regulatory approval, and the changing landscape of drug reimbursement. To mitigate some of these risks, companies within the industry have been and will continue to be active in forging creative and productive business relationships to bring new drugs to market.

The public often thinks of a pharmaceutical company as being fully integrated; however, the industry is in reality a network of companies working at times together and at times at odds in order to develop these new pharmaceutical agents. The network is complex and linked in that if one company is perturbed, say a contract manufacturing organization (CMO) or a large pharmaceutical company, there can be both direct and indirect consequences to the entire industry. Due to the networked nature of the pharmaceutical industry, there are a multitude of opportunities for innovative partnerships to emerge.

This chapter will examine several ideas on how collaboration within the industry promotes innovation. Pharmaceutical companies often add to product pipelines through licensing of new compounds, mergers and acquisitions (M&A), and in some cases collaborating with other pharmaceutical companies. Biotechnology companies traditionally are known for out-licensing compounds or technologies to large or midsize pharmaceutical companies and forming partnerships with similar-sized companies for drug development. The pharmaceutical industry has been quite successful at forging appropriate partnerships to access innovative compounds or technologies in order to continue to bring new medicines to market. Our challenge is to find new innovative approaches to working together within the pharmaceutical network in order to continue to make advancements in the development of new medicines.

Several traditional models of partnerships or business transactions have emerged within our industry to facilitate drug development and commercialization. We will review in brief these standard deal structures, while avoiding an exhaustive overview of past partnerships, and primarily focus attention on considering some innovative approaches to continue to advance drug development and commercialization through collaboration.

We then will turn our attention to two less recognized areas in which collaboration could lead to greater innovative opportunities: human capital of the research and development (R&D) organization and public education. Medical breakthroughs are made by scientists and physicians working in laboratories and the clinic to bring us the best treatments for disease. The pharmaceutical industry must examine the ongoing relationship with its R&D personnel and define innovative approaches to maximizing the ability of its scientists to engage in the cutting-edge research that is needed to develop the next wave of innovative medicines. In addition, the industry is known to have a transparency problem with the public. Long gone are the days when the primary care physician is the source of all medical knowledge and a patient faithfully takes a prescribed drug without questioning the choice. There is an onslaught of information readily available to the general public, the ultimate end users of key products. Some of this information is accurate; much is not. Media coverage of the industry focuses either on failure of a drug, the wrong-doings of a company, or, if positive, the overhyping of a breakthrough that in reality is years if not decades away from the market. It is no wonder that expectations for drugs are high and tolerance for the process of drug development is low. We will explore some ways in which community outreach and education may benefit not only the industry but also the overall wellness of society.

Looking then at how collaborations may facilitate the next wave of innovation, there are opportunities to reinvent the traditional models of collaboration between companies, within the organizational structure of research and development in a company, and with the beneficiaries of our products. Such innovation in partnerships could be a critical spring-board for the pharmaceutical industry to succeed in its quest to develop the next generation of innovative medicines.

Traditional partnering in the biopharmaceutical industry

The pharmaceutical industry is a complex network of companies engaging in different segments of the drug development and commercialization business. Each company has its own mission and vision, its own business model, its own set of products or services, and a cohort of interested parties within the company as well as outside investors (public and/or private) and customers to keep happy. Therefore, there is not a finite list of reasons why a company may or may not actively seek out partnerships to achieve its business goals. It is fair to assume though that a financial driver is behind a decision to partner.

If the company is small such as a biotechnology company or specialty pharmaceutical company, it is likely to be cash limited and resource limited. The small company may have embarked on a risky drug development program and could serve as an innovation center for new products, but this small company will reach a point when it lacks the expertise (e.g., personnel, equipment, manufacturing, marketing, sales) or the financial resources to continue its drug development program. The company may look to the industry network to find a partner to address its deficiency, for example, a large company to take over the costs of later stage drug development and/or commercialization, a CMO to supply compound, or a contract research organization (CRO) to engage in specific R&D activities. Such a partnership could strengthen the small company and allow it to pursue its innovative products or it could entangle the small company and limit its ability to remain at the forefront of innovative drug development.

Looking at a major pharmaceutical company, the reasons to partner are somewhat similar—filling a need that the company alone is not in a position to address, for example, a source of new compounds for development, capacity constraints on manufacturing, or a need to offset risk for a particular drug development program. Such large companies traditionally have been in a better position for negotiating collaborative arrangements due to financial strength. Many pharmaceutical companies, facing patent cliffs with their blockbuster products or lacking success with internal R&D efforts, have been under pressure to improve upon R&D and to fill what amounts to limited pipelines and so have been particularly aggressive in licensing clinical stage compounds for further development, and engaging in mergers and acquisitions of companies with promising pipelines.

For the sake of simplicity, here we will briefly review the partnerships that are traditionally entered into by either biotechnology companies or pharmaceutical companies when developing a new product. These are for the most part deals of potentially significant value to the partnering companies if all development, regulatory, and commercial goals are achieved for the drug product.

License agreements

The most straightforward type of partnering may be the license agreement. In a license agreement, one party (the licensor) grants certain rights to an asset under its control to a second party (the licensee). In the case of drug development, the license agreement usually addresses intellectual property rights, material rights, or both. This form of collaboration is basically a handoff in that the licensor steps aside and passes the baton to the licensee. The licensee controls further development and commercialization of the asset (e.g., a drug compound) according to a plan that has been approved by the licensor and is followed by the licensee. The asset can revert back to the licensor in the event that the licensee is no longer able to continue development, but in the case of termination of the license, the licensee no longer has decision-making authority over the asset. This form of partnering allows an organization that otherwise could not develop an asset into a marketable drug to find a company that has the capabilities to do so.

Licensors can be found throughout the industry network with many early-stage assets, especially platform technologies for drug discovery, originating in academia and government research laboratories. Early drug compounds developed by biotechnology companies may be licensed to pharmaceutical companies or larger biotechnology companies. For example, Isis Pharmaceuticals licensed an antisense drug to treat spinal muscular atrophy to Biogen Idec in 2012. Thrombogenics and Merck KGaA entered into a license agreement giving Merck the rights to a Thrombogenics drug to treat certain eye disease. Typically a drug candidate is licensed following completion of proof-of-concept studies, although some compounds may be licensed prior to clinical trials or after completion of a late stage clinical trial. A company might also license a compound that no longer meets the criteria of its pipeline strategy to another company for continued development, such as in the case of Pfizer out-licensing an early-stage cancer monoclonal antibody to VLST Corporation. Licensing is a mechanism for shifting the rights to develop drugs within the industry and ensures a means to initiate or continue development of drugs when the originator of the asset cannot or has a made a strategic decision not to retain a compound in its product portfolio. Innovation can be found in new opportunity rather than origin.

Collaboration or codevelopment/cocommercialization partnerships

Traditional collaboration within the industry involves two or more parties working together in drug development and commercialization. Collaborative partnerships allow companies to source complimentary drug development capabilities and share in the overall risk of the drug development process. Innovation can come from, among other things, cross-culture thinking and access to new capabilities or markets. Ideally, the two companies are equal in stature such that both are true partners in the collaboration with each sharing in profits and losses. Many successful partnerships of this type have been forged between biotechnology companies or between pharmaceutical companies. For example, AstraZeneca and Bristol-Myers Squibb Company established in 2007 a broad alliance to research, develop, and commercialize certain compounds for the treatment of type 2 diabetes and further extended the collaboration in 2012 to include the development and marketing of the diabetes product portfolio that Bristol-Myers Company acquired with its acquisition of Amylin Pharmaceuticals. A collaborative partnership is an excellent opportunity for innovation in that both parties can bring their respective expertise and resources to the collaboration, complimenting each other to create an optimized scenario for drug development. To maintain a successful partnership, companies should be committed, aligned in goals, and have decision-making authority that does not place one company at the mercy of another. Balance is key. The reality, however, is that a collaboration is often as strong as the weaker partner, and with the ever-changing landscape of drug development, any collaboration is subject to the pressures and goals governing each company individually. Collaborations are dissolved and in such cases one partner may be left with the burden of the decision on whether and how best to move the product forward. Shifting portfolio strategies within one of the partner companies may be responsible for the termination of the collaboration, such as in 2010 when Pfizer terminated its collaboration with Celldex Therapeutics for the development and commercialization of an experimental cancer vaccine. In many cases, unexpected negative data during clinical development may result in one or both parties opting out of the partnership. Each collaboration exists under a contract that attempts to address the potential ups and downs that might be faced over the life of the collaboration. In this regard, the collaboration approach is forward thinking in that it usually provides for contingencies in the face of clinical development setbacks or industry change.

Mergers and acquisitions (M&A)

Success or lack thereof can lead to a change in the number of players within the industry. In the case of a licensing arrangement or a collaborative partnership, success or failure of a drug development program(s) may lead to corporate reorganization of the involved companies in the form of either a merger or acquisition. Acquisition may be limited to a specific intellectual or material asset or may involve an entire company. When two companies have synergistic goals and capabilities, the preferred business model going forward may be to combine the two entities. Complimentary expertise or the desire to expand may be driving forces of a merger or acquisition. Common is the acquisition of a smaller company by a larger company, such as Takeda Pharmaceutical Company acquiring Millennium Pharmaceuticals in 2008 and Bristol-Myers Squibb Company acquiring Medarex in 2009. In many cases, the goal of the acquiring or surviving company is to access either a specific late stage or marketed product or overall pipeline of products. In recent years, the industry even has experienced the M&A of major pharmaceutical companies, such as Merck- Schering Plough and Pfizer-Wyeth.

In the case of the licensing and collaboration models outlined, success may lead the more financially secure partner to acquire the other in order to assume full ownership and control of a drug or drug pipeline. Assets and capabilities are both gained by the acquiring company or resultant merged entity and the organization now in control may be better positioned to develop and market the drug; however, there are unintended consequences of M&A that must be recognized. For example, the culture of innovation and risk taking typically associated with a smaller company can be lost as the acquired entity is integrated into the more complex organizational structure typical of an established multibillion-dollar corporation. Overlap in drug pipelines or therapeutic area programs could lead to mandatory divestiture of assets or discontinuation of programs that in the end decrease the overall number of compounds in development within the industry.

Contribution of partnering to date

So the historic partnerships of licensing or collaborating to develop a drug have been briefly outlined. The potential for M&A transactions resulting from one of these forms of collaborations or even as a stand-alone business transactions has also been introduced. The terms and the conditions of these transactions can be and have been in many cases quite creative thereby providing financial and business incentives for the pharmaceutical industry to continue to engage in collaborative relationships to advance drug development and commercialization. How then might the industry evaluate its ability and need to work in collaboration going forward in order to improve upon success to date and ensure that the approval and reimbursement processes within a changing landscape of drug development do not hinder the quest to develop the medicines of the future?

Innovative partnerships of the future

The pharmaceutical industry has been very successful at bringing new medicines to the public. Creative partnerships have enabled the industry to engage within itself to successfully supply the medical community with a bounty of medicines for the treatment of a host of diseases and disorders. Partnerships have driven the development of both specific medicines as well as platform technologies to discover and create new classes of medicines, and thus form the foundation upon which the industry has had its successes. In the last year alone (2012), there were over 30 new molecular entities approved to treat diseases and disorders ranging from basal cell carcinoma to severe hypertriglyceridemia to cystic fibrosis. The pharmaceutical industry continues to be vital to the ongoing quest to maintain and improve upon the quality of human life.

Despite its contribution to the overall wellness of society, the pharmaceutical industry is under constant pressure from the public to bring novel medicines to market ever faster and ever cheaper. Companies look to streamline operations and find new measures for cost reduction. The regulatory and reimbursement environments have changed in response to public pressure on the industry to create safer and more efficacious medicines at low cost. Business as usual for the industry is no longer tenable. The pharmaceutical industry must adapt to its changing environment and identify innovative approaches to drug development and commercialization if it is to survive and thrive in the future. Creative partnering within the industry and with customers is one critical component to achieving success in the future.

Industry-wide collaboration or the megacollaboration

There are several common challenges to the drug development process that every company must find a way to address. As reviewed in other chapters, clinical development challenges include, but are not limited to, the need for a universally accepted data entry and management system;

the requirement for standardized criteria for clinical trial site qualification and training; and an overall reduction in the cost of clinical studies. Bringing pharmaceutical and biotechnology companies together to share best practices, identify common challenges, and jointly define solutions seems to be an obvious area where collaboration can succeed. In the autumn of 2012, ten biopharmaceutical companies announced the formation of the new nonprofit organization—TransCelerate BioPharma, Inc. With a who's-who list of pharmaceutical entities as members of this nonprofit organization, benefits from cross-culture thinking and collaborative efforts should allow for the industry to tackle the challenges that all companies encounter with respect to clinical development studies. It is a bold initiative and one that with true commitment should yield results that reduce development costs across the industry. It will be interesting to see what solutions may be found by this organization and whether indeed ten companies can come together under one umbrella organization for the common good of the pharmaceutical industry.

Presuming success of a partnership such as the TransCelerate initiative at least as a starting model, the industry should be able to define other areas of common interest whereby banding together as one may benefit the industry overall while not compromising the direct competitiveness of the pharmaceutical market. Although improvement in the drug development process and the reduction in the costs mainly associated with clinical trials will indeed lead to efficiencies within the industry and perhaps better balance sheets for individual companies, these activities will not necessarily improve upon our ability to discover new drugs and engage in the early clinical studies necessary to support advanced clinical development. Innovation is needed across the network to improve upon the overall process of drug development, but innovation in pharmaceutical agents will only occur if we find ways to revamp the methods for drug discovery and early clinical development.

The industry should form collaborative entities, or megacollaborations, to support early research and development of new molecular entities and biopharmaceuticals. Perhaps those ten biopharmaceutical companies or another cohort of companies can come together as members of a nonprofit institution that is tasked specifically with discovering new compounds. To simplify, perhaps a few intractable disease areas could be the focus, for example, Alzheimer's disease, aging, or infectious diseases. The new collaborative organization could be financed, for example, on an annual basis by each member company in equal installments so that no one company is the "lead" party, would have its own governance structure, and would have an annual plan and budget voted on by the founding member companies. Product development programs would have plans and milestones set and approved by the board (one representative per founding biopharmaceutical company). Upon successful completion of defined milestones along the development path, the member companies would have the opportunity to license the compound from the collaborative entity for later stage development and commercialization. License terms would be predefined by the board for each compound prior to initiation of clinical studies. Once licensed, the licensee would pay milestones and royalties back to the nonprofit organization to be used to fund new or ongoing programs. Basically, the pharmaceutical companies that are members of this nonprofit precompetitive R&D organization would have a discovery engine that would churn out new investigative compounds for the benefit of the entire industry. The nonprofit could hire personnel directly or could operate with employees from the member companies. The organizational structure should be modular and would need to be revised based on regulations. The industry and its regulators should be able to define approaches to generate not just one but a few specific legal entities that are disease-area specific and a reflection of the cooperation of the industry at large. It is a simple idea where the devil is in the details. The industry must transform its traditional sources of new molecular entities or biopharmaceuticals, and jointly work to offset the risks of early clinical development of these assets since the current model of drug discovery and development is not yielding innovative compounds at the pace required to meet our health needs of the future. Similar collective initiatives potentially could streamline other stages of drug development, such as manufacturing, safety testing, and regulatory agency interactions.

Increased collaboration between diagnostic and therapeutic companies

Drug discovery and development are increasingly complex. Perhaps the increasing knowledge base of human biology emboldens us to attack more complex problems. Perhaps the industry already tackled the easy and obvious biological challenges, and currently lacks the knowledge to address the more complex diseases and disorders. Diagnosis of a particular disease or disorder is the first step in treatment. Personalized medicine or directed therapeutics has emerged as a new paradigm in drug development, as the understandings of the molecular basis and physiological manifestations of disease have led to more accurate measures for diagnosis. The pharmaceutical industry has begun to realize that increased collaboration between diagnostic developers and drug makers could lead to improvements in the traditional drug development process. For example, Pfizer in 2011 partnered with Qiagen to develop a companion molecular diagnostic test to identify the subset of non-small cell lung cancer patients most likely to respond to a Pfizer investigational compound in clinical development. Eli Lilly acquired Avid Radiopharmaceuticals in late 2010 to access a diagnostics development platform covering disease areas such as Alzheimer's disease, Parkinson's disease, and diabetes. Collaborations between pharmaceutical and diagnostic companies will continue to be established. How then might the partnering strategies of these two sectors of the industry evolve in support of greater pharmaceutical innovation?

Diagnostic companies, in particular molecular diagnostic companies, are beginning to make significant contributions to the overall drug development process and influence the pharmaceutical markets. Pharmaceutical companies are under increased pressure to develop targeted drugs, those compounds that knowingly benefit a specific patient population. Payers, prescribers, and patients want to know ahead of treatment that the prescribed drug is likely to have a therapeutic effect with limited side effects. The field of molecular diagnostics is developing products to identify those targeted patient populations. Now during the course of drug development, clinical data is generated both in responders and patients not responding to treatment. This clinical data can be useful in support of the drug development plan as well as in support of the development of diagnostic agents for a patient population. Drug companies and diagnostic developers should explore collaborative opportunities early in the development process to jointly develop therapeutics and diagnostic agents that specifically identify those patients whom would benefit the greatest from those therapeutics.

What if in the end the diagnostic product is the key development that enabled the drug maker to gain market approval in a specific patient population? What if postapproval monitoring of the drug with the diagnostic agent has allowed for market growth? What if the diagnostic agent actually helped to identify the correct population in which to conduct the initial clinical trials, thereby reigning in clinical trial costs by focusing development efforts in those therapeutic areas where patients are most likely to respond to the drug? The value of the contribution of the diagnostic agent and the role of the molecular diagnostic company in the overall development of the drug may need to be reexamined. Currently, molecular diagnostic companies capture only a portion of the value that their products contribute in drug development and commercialization. Collaboration occurs but actual true value contributions of the parties may not be represented in the deal terms. Pharmaceutical companies still capture the majority of the upside of drugs on the market. Perhaps drug companies should look at diagnostic companies with less disparity in value contribution and explore new avenues of collaboration that would benefit both parties while leading to safer and more specifically efficacious drugs for a given patient population. This idea can be carried through to other players within the pharmaceutical network, in that if we can find a way to shift the parity in many deal structures both with upside and with risk, perhaps in partnership more drugs will be able to advance successfully through development and to market.

Partnering with academic institutions

Both government research entities and academic institutions have been instrumental in the success of the industry. Scientific discoveries abound outside of the confines of the corporate world. The driving force is different. Government and academic institutions are nonprofit entities. Both exist to serve the public and both are sources of innovative science. Talented researchers seek careers at government or university laboratories to maintain the quest for scientific knowledge and a flexibility to take risks in pushing the boundaries for scientific discovery. The biotechnology industry was born from the desire to find avenues to commercialize inventions made within the walls of academia. How then can we continue to foster the fundamental relationships between the pharmaceutical industry and the scientists at these institutions? How might those in the business of creating new medicines redefine the traditional interactions between industry, government, and academia to provide for a more innovative environment for the future of drug development?

The pharmaceutical industry has begun seeking new relationships with the academic world and has recognized new opportunities to expand or establish collaborations. Many pharmaceutical companies as well as mid- to large-size biotechnology companies have located or established new sites in what are considered some of the hotspots of biomedical innovation, including, Boston-Cambridge, Bay Area, San Diego, and internationally in Europe and Asia. Johnson & Johnson announced in the fall of 2012 the establishment of several "innovation centers" to bring in close proximity some of its researchers with leading academic research centers. Pfizer, Merck, and Novartis have research centers in Cambridge, Massachusetts, home to a multitude of biotechnology companies spun out of this region's academic centers. Proximity of these industry scientists to scientists at leading academic centers will promote scientific discussion, enable innovative ideas to be shared, and can lead to new approaches in drug discovery and development.

Breaking down the barriers of communication and reexamining those business restrictions that each type of organization places on its researchers are key to maximizing the value of such relationships. Pharmaceutical companies have it right in fostering interaction with the life science communities that in many cases serve as the sources of medical innovations. Promoting collaboration between industry and academia at an early stage with industry truly educating the research community on the process of drug development, especially clinical development and regulatory science, could lead to more mature innovative ideas stemming from both industry and academic research laboratories, ones that the pharmaceutical industry can be a part of nurturing early to yield realistic product opportunities.

By placing R&D centers near academic centers of excellence, new avenues for collaboration will emerge. How will funding of these new collaborative initiatives occur? Traditionally, a company would pay or "sponsor" research conducted in a university laboratory, usually research conducted by a single lead academic scientist under a specific project plan. The sponsoring company then had a right to license any intellectual property developed during the course of the sponsored research project. This model may not be optimal for supporting innovative research of the future. As pharmaceutical companies locate or relocate to the hotspots of biomedical innovation, opportunities emerge for more involved collaborations that extend beyond single sponsored research programs.

Today the academic scientist is under considerable pressure to find alternative sources to government funding in order to support research programs. Greater access to industry funding of programs may be a solution for the academic scientist to the benefit of the industry. Pharmaceutical companies in recent years have actively pursued broad disease-specific or technology-platform-specific alliances with academic centers. Novartis and the University of Pennsylvania in 2012 entered into an alliance to develop personalized immunotherapies based on a certain technology developed at the university. The alliance of scientists at the University of California, San Francisco with Sanofi researchers to develop new drugs to treat diabetes is another example of a broad, disease-focused collaboration. More broad collaborations such as these wherein pharmaceutical companies can provide drug screening, clinical development expertise, and financial support to academic scientists are needed.

Biopharmaceutical companies overall should be more aggressive in funding biomedical research and translational medicine at academic institutions or at early-stage companies that are founded by academic scientists. The early scientific understandings of today can be the breakthrough products of the future, but will only realize full potential with solid financial backing. Pharmaceutical companies are positioned with substantial cash resources that could be employed more strategically to fund research at academic centers in a collaborative manner. Similar to the idea of the collaborative R&D nonprofit organization outlined earlier, either an established syndicate of pharmaceutical corporate investors or a new entity comprised of multiple biopharmaceutical companies could be established that would focus on the future for the industry as a whole in certain disease fields or technologies. Companies could collaborate to collectively fund early research in areas such as regenerative medicine and neurodegenerative diseases at academic research centers. Industry is working to establish such ventures. In 2012 seven biopharmaceutical companies partnered to establish the Massachusetts Neuroscience Consortium with the purpose of jointly funding preclinical neuroscience research at academic research institutions in the state. Presuming success of this initiative, perhaps additional consortiums will be founded globally to fund preclinical research in other disease areas as well as to fund the development of technology platforms that could serve to generate new pharmaceutical agents. Seeding the garden of biomedical opportunities in academia and with other nonprofit institutions will benefit the industry at large as it looks at its current product pipelines and questions the sources of the innovative medicines of the future.

Innovative collaborations with government research institutes

The federal government recognizes the need for innovative collaborations with industry to overcome the slowdown in drug development. The National Center for Advancing Translational Sciences (NCATS) was established in late 2011 as the newest center at the National Institutes of Health (NIH) with a focus on improving translational science in order to accelerate the development of new therapeutics and diagnostics. NCATS has engaged industry in discussions early on to identify new ways of cooperation. Initially eight pharmaceutical companies agreed to participate in a pilot discovery program by providing validated compounds for testing in new therapeutic settings. NCATS agreed to contribute its preclinical and clinical capabilities across multiple institutions. Collaborating with the pharmaceutical companies, NCATS may be able to uncover new uses for the compounds provided by industry partners. With the companies maintaining ownership of their compounds, this is a success for all involved in that new science and potential medical advancements can be made while new markets may be uncovered for the industry. With government resources combined with industry knowledge, the drug development process may be accelerated at least for compounds that already have been validated by industry in one therapeutic area but otherwise might not be developed in other therapeutic areas due to the business concerns of a company (e.g., market size, budget constraints, or lack of development resources.) Hopefully, the NCATS initiative will be successful and serve as a model for continuing collaboration between the pharmaceutical industry and government laboratories. This case illustrates only one possibility of the potential organizational relationships that could align public and private interests in advancing the development of pharmaceutical agents.

   As with partnering within the corporate network, future collaborations between academic or government research institutions and biopharmaceutical companies may need to be defined on new business terms, terms that better reflect the roles of each party in the development of a new drug. Both sides should revisit current practices and identify ways to be a bit more flexible on the standard business issues, such as intellectual property, confidentiality, liability, and financial terms. The pharmaceutical industry should continue to explore new collaborative structures with government and academic research centers to the benefit of medical science and the industry at large.

Collaborating with research and development groups to promote innovation

Defining and implementing new collaborative relationships within the pharmaceutical network and with network associates, such as academic research centers and government laboratories, could improve upon innovation in drug development. In this section, we will explore opportunities for the industry to increase collaboration by promoting an innovative culture within each pharmaceutical company. In particular, this section will examine how a company can collaborate with its innovators to enhance the drug development process.

Innovation leading to new product opportunities emerges during both drug discovery and clinical development. Thus, the research and development teams that drive these activities within a company, the ones tasked with tackling the complex biology of the human body, are the engines for developing new medicines. Increasingly it has become more difficult for these scientists, engineers, and clinicians to thrive. Consolidation within the industry driven by the economic realities of current times has threatened the job security of these innovators, which in turn hinders the necessary motivation to push innovation. Innovators benefit from supportive work environments when embarking on the discovery and development of the next generation of medical breakthroughs. Corporate structures have become less fluid such that collaboration across R&D divisions can be limited, with researchers limited to or "siloed" into specific projects. Further, the best and brightest researchers are commonly tracked into management roles, which may be good for career development but could result in a loss of intellectual capital on the discovery and development end. These combined pressures of lack of job security, limited collaboration across R&D teams, and loss of the top scientists to managerial responsibilities work in concert to limit the innovative potential within a company. By addressing these issues, the industry may promote an environment in which enhanced collaboration within the R&D group can lead to increased innovation.

Security to innovate

Innovation requires the ability and willingness to accept risk. Risk associated with product development is regulated; however, business risk, that is, the willingness to fund a therapeutic area or maintain an active R&D team, is a corporate decision made in many cases not by the scientists and clinicians but by the lawyers and business professionals in the company. The majority of those working within the pharmaceutical industry chose this path due to a strong belief in science, medicine, and the ability to improve human health. These industry veterans include those that are not directly involved in research and development. Business realities must be taken into consideration, but the industry could benefit by reexamining the purpose of its products in society and then defining new collaborative environments in which R&D innovators and their business comrades better align interests when pursuing medical advancements.

Pharmaceutical companies should redefine their collaborative environments with an emphasis on creating cultures of innovation where constant worries of job security are minimized. Although the industry is being squeezed, the medical needs of the human population are ever present. Companies need to invest in the future of drug development by reducing the uncertainties rampant in the industry with respect to job security. Less emphasis on workforce reduction and corporate restructuring initiatives should be embraced. It is interesting that entrepreneurs are rewarded for the experience of starting new ventures whether their bets are successful or not, but industry scientists are often penalized based on whether their bets generated a new drug. This paradigm does not promote innovation. With downsizing in the industry being as extensive as it has been, companies should now support the current R&D workforce and rebuild or renew support of innovative discovery and development cultures. Perhaps an industry-wide moratorium on R&D restructuring should be initiated for the next decade to allow researchers and clinicians the time to regain their confidence in science and champion the risky, early-stage programs necessary to lead the advancement of modern medicine.

Greater collaboration across the research and development (R&D) organization

The structure of the R&D organizations of today threaten to impede the passion or innovative spirit necessary in order to tackle the complexity of human biology and defy nature with our novel therapeutic approaches. Innovation ideally occurs when R&D teams can do what they do best: think, read, and look at data as well as interact with colleagues to explore and challenge ideas to define a path forward in drug development. Pharmaceutical companies can collaborate with other companies, with academic organizations, and with government research entities to promote new collaborative initiatives. Internal politics however can make it challenging for different R&D divisions within a company to work together. Biopharmaceutical companies should foster collaboration among the innovators within its organization, across research divisions, and with patient groups in order to provide the energy boost needed to spur the discovery and development of new drugs.

Companies need to break down the internal barriers of communication that act as roadblocks to transferring innovations across disease areas. Pharmaceutical companies traditionally organize R&D efforts around therapeutic areas or particular drug development programs. In many cases, these R&D groups compete for company resources and the survival of their programs, not to mention jobs. Focused efforts are needed, but companies should establish collaborative environments that encourage data sharing and introduce new initiatives for brainstorming across therapeutic areas. Perhaps a drug being developed for cardiovascular disease may actually have anti-inflammatory properties beneficial for rheumatoid arthritis treatment. Clinical data may yield insights into how a drug might be useful in lowering cholesterol rather than its initial purpose of treating asthma. Drugs can be repurposed for new indications. Promoting more interactions across an R&D organization may lead to new breakthroughs. Management support of data sharing and collaboration is essential so that scientists are encouraged to work across the organization without the threat of penalty should the development plan for a drug change. The silo approach to drug development needs to change so that the walls imposed by an organization do not impede its potential to development the best new treatments.

Additionally, companies should promote the innovation mindset by reminding employees throughout the organization that patients do benefit from drugs. Programs to connect R&D professionals, especially those not directly participating in clinical development, with patient groups should be established. Patient populations should be engaged to educate potential innovators as to the realities of living with disease. Experiencing what it means to have an untreatable disorder can motivate and inspire innovation. Further, specific knowledge of a disease state gained directly from those suffering, not just from journal articles and textbooks, may yield new understandings and ideas for treatment options. There is no better expert than the patient living with a disease or disorder. Collaborating with patient populations to educate us about the true reality of disease can and will motivate us for future advancements in medicine. Scientists and clinicians should not lose sight of why they decided to embark on a career in the pharmaceutical industry. Companies should embrace collaborative interactions with patient advocacy groups and not penalize its innovators for forward thinking developed through this outreach, but instead reward them for making strides to define new medicines by working with those in need.

Time for innovation

Medical breakthroughs take time and focus to achieve. Today's thought leaders, those within a company that come up with the new ideas and approaches for drug development, are often scientists or physicians who have limited time to focus on the next breakthrough. Many of the pharmaceutical industry's research and development leaders instead spend a majority of their days in meetings and may not have the time to directly engage in innovative activity.

Meeting commitments increase as an organization grows in order to coordinate the ever more complex organizational hierarchies. Corporate culture codifies decision-making processes, and collaboration between business and science leaders within an organization governs corporate success. As such, the time that R&D leaders spend in meetings has increased as organizational complexity has grown, and this time potentially could have more overall impact if spent focusing on innovative science. Perhaps one characteristic that separates the early start-up energy of innovation from the large company process mentality is the ability of R&D leaders to remain directly involved in development programs on a daily basis.

Why do we sequester the most talented R&D scientists in an endless string of meetings? Such innovators have most likely been successful in the laboratory, have risen in the ranks of management, and therefore are tasked to be a part of or at least present for every decision (or decision to postpone a decision). While there are key decisions that scientists need to make or contribute to making, the industry would benefit from defining a new collaborative working environment that allows each company's best and brightest innovators to do what they do best–explore and capitalize on scientific understandings in order to develop the most cutting edge human therapies. The enlightenments of these thought leaders do not occur in the constructs of fully scheduled days. Instead, breakthroughs occur when scientists can be scientists and spend more time thinking, reading, looking at data, and interacting with colleagues.

Pharmaceutical companies should define new collaborative structures from within that will honor the capabilities of bright scientists, physicians, and engineers while allowing for the business component of drug development to continue. It is a daunting task as the experience of these senior scientists makes them valuable when bridging the business and scientific sides of drug development. Yet, leading scientists are stretched too thin. Business leaders need to work with scientists to find better ways to organize the corporate environment so that R&D leaders can directly engage in scientific discovery. As one example, two individuals could share a senior position in R&D so that the duties of management and scientific leadership are split. The two individuals would rotate monthly or annually between managerial and scientific responsibilities. Job performance would be the result of a collaborative approach to fulfilling the responsibilities of the role. In the end, each individual on the job "team" would have the opportunity to remain closer to the drug development programs while still contributing to management decisions and corporate leadership. Limiting the time that these thought leaders spend addressing issues not directly related to the develop of specific compounds may enable a company to capitalize on the innovative natures of its senior R&D leaders to better focus on the key mission of the company—the development of new therapeutic products. This endeavor will involve collaboration across the many divisions of the company and will need to have the official support of senior management and the corporate board to be effective. Greater commitment to science rather than the organizational process should be emphasized in an innovative organization.

Collaborating with the public

In addition to collaborations on the scientific front to promote innovation, the pharmaceutical industry may benefit from establishing a new relationship with the end users of its products: the public. Pharmaceuticals are difficult to understand in general and the pharmaceutical industry lacks transparency. As a result the industry is not well understood. The public often accuses the industry of taking too long to develop new drugs, and when there is a new drug, the cost to access is too high. On an experiential level, a doctor prescribes a drug (or a visit to the local pharmacy leads to a purchase of an over-the-counter [OTC] product); the patient has to pay (and in many cases thinks the drug costs too much); and the drug works, or does not work, or has some side effects. Because the patient followed prescribing directions and naturally expects that the drug should work for everyone, when it does not work, the patient can only blame the drug maker. Such unrealistic expectations arise from a fundamental lack of understanding of the process by which drugs are developed and a fickle appreciation for the ultimate benefit that these products can provide.

In order to create the breakthrough drugs of the future, the industry will need strong support from the general public. The public needs to be reminded that the industry is engaged in the development of life-altering medicines. Pharmaceutical products are not for entertainment; not for communication purposes; and not for standard life needs such as clean water, roof above head, food, and clothing. Pharmaceutical products improve or extend the quality of life. Importantly, it must be recognized that to have such interventional power over human biology, these products are not risk-free. Even so, most will agree that, given the prospect of watching a loved one die in pain or live decades in agony, the availability of a medicine to treat the underlying ailment is a miraculous gift. And a miracle it truly is considering how difficult it is to develop and receive approval of a new compound. How then can we ensure that the public is willing to support the continued advancements in medicine that new pharmaceuticals contribute and is ready to accept the inherent risks and costs associated with therapeutic products?

Engaging the public through a variety of mechanisms to enhance mutual understanding and appreciation could significantly impact the public perceptions and discourse around drug development. The industry needs to work on its outreach and more effectively educate the general public as to what it takes scientifically and financially to discover and develop new drugs. Collaboration with the public—the industry's end users—can strengthen the pharmaceutical industry's innovative potential. Better understanding of the drug development process could lead to better appreciation for the products that we provide and inspire additional support for the industry at large. Expectations for the drug development process need to be managed, and understanding of the process—both business and science—is key. Otherwise, the desire for cheap, cutting-edge, and risk-free medicines will hinder future pharmaceutical advancements by leading to increased regulation of and penalties to industry players when the public's unrealistic demands are not met.

Further, the public is in reality the source of the industry's future innovators. Potential innovators should not have to wait until graduate school or their first jobs to finally understand what it takes to develop a new medicine. Sometimes the naivety of our youth can provide the source of creativity to address an otherwise intractable scientific problem. This may hold true in pharmaceutical science as well. Direct collaboration with educational institutions to provide real-world learning of the drug development process can benefit the industry by stimulating the thought leaders of the future and encouraging them to pursue careers in the pharmaceutical sciences. There should be an increase in industry-sponsored training within academia both at the graduate and undergraduate levels. Classes in the process of drug development and regulatory science should be taught by R&D folks employed at a biopharmaceutical company, not just university professors, many of whom have never worked in the pharmaceutical industry. True knowledge transfer of the process of developing a drug is necessary to motivate and inspire the next generation R&D teams.

The public also includes the patient populations that experience the very diseases the industry is attempting to alter. By reaching out to patient populations, pharmaceutical scientists also can gain better understanding of disease and perhaps more important stimulate an increased passion within the industry to find treatments to the most challenging of physical ailments. Greater communication with patients could allow more researchers to gain direct knowledge to apply to drug discovery.

Working with patient advocacy groups could even lead to political support for championing the drug development process.

Collaboration with the public for greater understanding of drug development will benefit the industry overall. An industry-wide consortium with a mission of education should be established. Focus should never be on a particular product but instead on the overall process of drug development from discovery, to regulatory approval, to market. Confidentiality concerns aside, more can be gained by teaching others how the industry operates rather than maintaining the veil of secrecy that the industry has tried to maintain. Who knows, the next big development program to pursue could come from your neighbor who likes to experiment in his or her garden or your grandmother who remembers what her grandmother did to treat the flu. Educating the public will stimulate more minds to contribute to and more hearts to appreciate the mission to identify new therapeutic agents to maintain and improve upon human health.

Conclusion

Opportunities abound to use collaborations to promote innovation throughout the industry. The industry is a network of companies that has been and will continue to be successful in establishing partnerships to advance the development of new drugs. New initiatives to promote greater collaboration within the industry and with external stakeholders will be a key to ensuring future successes. Industry-wide collaborations or megacollaborations targeting a disease area may lead to an improvement in drug discovery and development by increasing intellectual capital while reducing costs and minimizing risk. Enhanced collaborations with academic research institutions and government research laboratories may provide for new avenues of learning and allow for the inclusion of new thought leaders in the quest to develop new breakthroughs in medicine. At the heart of pharmaceutical innovation is the research and development organization. Support of R&D organizations throughout the industry in the forms of greater job security, increased allowance for collaborations across research divisions within a company, and time-management programs that increase focus on actual drug development should strengthen the innovative spirit within the pharmaceutical industry that will lead to the breakthroughs for tomorrow's medicines. Educating the industry's potential innovators of the challenges of today's patient populations and in turn educating the general public as to the processes by which new medicines are made available will encourage continued support, passion, and the resolve necessary to meet the health challenges of today and tomorrow.

chapter three

Portfolio management

Selecting the right "set of projects" to meet product development goals

Arun Kejariwal

Contents

Figure 3.1 Big picture fit for portfolio management.

Introduction

The objective of this chapter is to share the importance of portfolio management concepts, especially in a resource constraint environment, in meeting a company's strategic goals. In particular, the chapter focuses on the following key areas:

- Overview of portfolio management and what it entails
- What is needed to optimally align the portfolio to deliver on strategic and operational goals
- Best practices and innovative approaches to portfolio management
- Recommendations on frequently encountered challenges with management of a portfolio of drug development projects

In the context of this chapter, a portfolio is defined as a collection of assets and projects that are in the preclinical to registration phase of development, including line extension and enhancements to products that are already on the market.

What is research and development (R&D) portfolio management?

Research and development (R&D) portfolio management is a systemic approach to choosing, monitoring, and allocating resources to those assets that when brought together maximize the opportunity to meet or exceed a company's strategic goals. Portfolio management utilizes project-level assessments and alignment with strategic goals in selection of a portfolio with maximized value and manageable risks.

Portfolio management starts with optimizing what is in the portfolio and then expands into defining the portfolio, that is, what should be feeding into the portfolio from basic research and in licensing.

Portfolio-level analysis can also help set realistic strategic goals and inform various execution options for meeting those goals. For example, portfolio analysis can be used to understand the implications of playing everywhere versus playing in core-focused areas.

Why is portfolio management required?

History suggests that the pharma model with an uncontrolled R&D budget does not seem to be working. As a result of declining productivity, the R&D funding has been constrained either because of the self-imposed limit on R&D spending or just because of limited available funds. The healthcare ecosystem and its needs have also evolved. Good science alone is not sufficient to have a meaningful health and economic impact on the society. Regulators are increasingly cautious and require data showing clear differentiation from existing treatment. Insurances and payers demand a strong value proposition for the drug to manage the exponentially rising cost of medicines and healthcare. This particularly adds another and somewhat new dimension to the drug development challenge.

Incorporation of a consistent and systematic portfolio management approach with the right discipline provides the needed focus and alignment to help improve return on R&D investment. Portfolio management adds value through better insight, improved decision making, and increased alternatives.

What characteristics of the R&D portfolio need to be managed?

Portfolio risk profile

Which portfolio protects the organization against unfavorable events such as clinical trial failure, shifts in the marketplace, or insufficient resources? What is the trade-off between risk and expected return? Is the overall risk of missing aspirations tolerable? Is there transparency and alignment in the organization of this risk? For example, choosing portfolio C over portfolios A and B sacrifices some expected value in order to gain greater confidence of achieving expected value and to reduce the risk of not meeting targets.

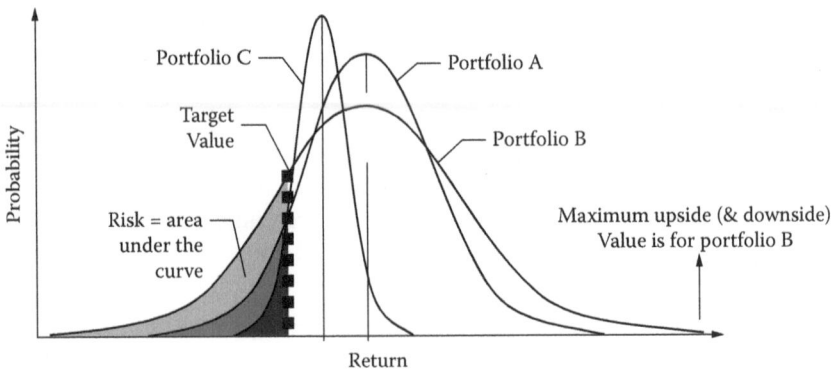

Figure 3.2 Risk profile for different portfolio choices.

Sources of risk

Risks in drug development are dominated by two major sources of uncertainty:

1. Uncertainty that a compound has approvable and marketable safety, efficacy, and therapeutic properties—regulatory and technical risk
2. Uncertainty that the market will support the expected price and volume of the approved compound—commercialization risk

Additional uncertainties are associated with cost; timeline; and macroeconomic, political, and regulatory phenomena.

Risk correlation and impact

The greater the positive correlation between multiple projects in a portfolio (i.e., common events that subject all projects to failure), the higher the risk. Hence individual investments must not be chosen purely on an individual basis but rather on how they interact with each other.

Compound correlations introduce risk—Multiple investments in compounds fail to produce, drugs as compounds share a similar fatal flaw (e.g., structure, binding, and covalent bonding).

Therapeutic correlations introduce risk—Multiple investments in compounds of the same class or disease fail to produce drugs as the mechanism of action may prove to be ineffective or unsafe.

Market correlations introduce risk—Compounds and diseases share competitive threats (e.g., changes in pricing or reimbursement policy for a specific disease or class of compounds).

Breadth versus depth, aka diversification versus focus

A focused investment in one mechanism or disease area increases the probability that at least one candidate will be successful. However, correlation increases the risk that the investment will yield less than expected and potentially no return.

A diversified investment across each noncorrelated mechanism or disease area minimizes the probability of zero products. However, diversification will decrease the likelihood that a candidate from the "preferred" mechanism or disease area will be successful and can lower the expected value of a portfolio.

Alignment with top-level organization strategy

The portfolio choices should inform and enable business strategy. Formally and regularly measuring portfolio performance against explicitly articulated objectives that derive from the business strategy is essential to ensure continuing alignment.

Unique aspects of portfolio management in pharmaceutical industry

The core objectives of the portfolio management across industries is the same, that is, to select a portfolio that maximizes the ability to meet the company's strategic goals. The following industry specific project attributes and associated investment and industry environment provide unique challenges for pharmaceutical portfolio management.

Long duration between start of investment and realization of benefits— Projects that get added to the portfolio today will potentially add to the portfolio value in 10 to 15 years.

Only a very small portion of all projects yield benefits—Historically, less than 5% of the projects entering clinical development make it to the market. Moreover, the intrinsic value of any failed investment or project is minimal to zero.

High cost of getting a drug to market—The total cost of getting a drug to market runs in the $1 billion range.

Intellectual property protection at compound level and not at mechanism of action (MoA)/technology level—The patented product life continues to get shorter due to longer cycle time and hypercompetitive markets.

Investments are not fluid—R&D projects are not easily traded as the financial investments. Unlike financial or other investments, there is large cost associated with closing one R&D project and switching to another.

Investment in a partial asset is not a normal practice—Unlike financial investments it is not possible to only partially invest in an asset. The joint venture and partnership model in drug development do provide some opportunity for partial investment in an asset.

High "inertia" for investment in projects moving to late stage—There is a high level of pressure (as a result of internal culture or Wall Street implications) to continue to explore every option, regardless of the likelihood of success, before terminating development of a late stage asset.

Supply cannot be restricted to manage price—Unlike some industries (for example, OPEC restricting oil production per year) it is unethical and illegal to intentionally lower production of medicine to artificially lower supply to influence the price of a drug.

Highly regulatory environment—The testing requirements to demonstrate clinical value of a compound for approval to market the drug is controlled by regulations and governmental bodies.

End consumers rarely pay the full price of a drug—In most healthcare systems, insurance companies or their delegates pay for the medicine. The end consumer, if insured, pays the premium to the insurance company and perhaps a copay. Pharmaceutical companies have to negotiate the drug price with the payers and the "value" of the product is rarely determined by the consumer/patient.

In spite of these unique challenges, one can and should draw parallels and learn from the portfolio management practices in other industries.

Portfolio management prerequisites

Organization readiness and expectations

Portfolio management requires another layer of complexity, one that seems counter to the culture within pharmaceutical companies, since most activities are oriented toward the individual projects and not the entire R&D portfolio. The foundation of portfolio management entails allocation of limited resources to the set of projects that enable progress toward company strategy and not that of just the individual project or business area. Organizations should have the culture and incentives in place that promote this thinking and ensuring that the leadership is committed to account for both short- and long-term implications of their decisions.

This means that each business leader should have the maturity and incentive to give more importance to the value of the company portfolio and not just focus on ensuring funding of the projects within their own business. Similar behavior should also trickle down to the therapeutic area (TA) leads within each business, that is, they should be willing to give up resources to the better investment opportunity in another TA.

The leaders in the organization should be savvy with the portfolio management concepts and its application in the pharmaceutical industry. Training should be routinely provided to ensure sincere engagement and a common understanding of the terminology, value metrics, approach, and methodology.

There should be a good balance between data-driven and gut-instinct decisions. Gut-instinct-based inputs should be clearly articulated (rather than "I said so") and factored into the decision making.

A transparent and constructive debate during decision making should be encouraged and is sign of a high-performing portfolio management process.

Transparent process, accountability, and decision ownership

A standardized portfolio management process with clear delineation of roles and responsibilities, accountability, decision ownership, and decision communication plan is essential for successful implementation of portfolio management. There should be checks in place to ensure everyone impacted (directly or indirectly) with the portfolio decisions has visibility and buy-in (prevent "black box" thinking) on the approach and the methodology being employed for the portfolio decisions.

There should be a consistent framework across the organization for project-level inputs, assessment, and decision making. The portfolio management process, among other things, should include project valuation, uncertainty assessment at the project and portfolio level, project prioritization, portfolio balance, and portfolio optimization. It should include selection of a portfolio along with multiple "fallback" portfolios.

It is best practice for a "neutral" group to own and facilitate the process. It can inject an objective perspective and encourage transparent and constructive debate during the decision-making process.

The sophistication of the portfolio management process and associated tools and methodology should closely mimic maturity of the organization culture and readiness for portfolio-type thinking, that is, either one should not be too far ahead of the other.

Portfolio management should also consider the execution challenges and include a framework for consistent assessment of factors at the project level that impact the portfolio goals.

Translation of company strategy into portfolio goals and measures

The company strategy sets the "what" and portfolio strategy sets the "how." The portfolio management team should have a clear understanding of the company strategy and be able to succinctly articulate it to the wider audience throughout the portfolio decision process.

The portfolio metrics and targets for decision making should align with the company's strategy priorities (for example, maximize the value to the shareholder and/or society, maximize revenues in near term, balance number of projects across the pipeline, and/or maintain high single-digit revenue growth). The measures, their priorities, and targets should be clearly defined and agreed upon by the leadership across the organization. The portfolio metrics and targets are unique to a company need and the stage of evolution in its business lifecycle.

Any critical success factors (leading and lagging indications) should also be identified for ongoing progress monitoring.

Alignment on uncertainties that impact portfolio goals

The key uncertainties that could impact the strategic objectives of the portfolio should be identified, agreed upon by the leadership team, and documented for ongoing monitoring and impact assessment.

These uncertainties should be segregated by their origin and impact, for example uncertainties at project level, at TA level, at function level, at organization level, at industry level, and at cross-industry level (Table 3.1).

There should be clear articulation of how and when the impact of each of the uncertainties will be modeled to understand their implications on the portfolio metrics. This is essential to prevent impact assessment of the same uncertainties at multiple levels.

Understanding of any interdependencies (positive or negative) between the uncertainties will provide for a more robust portrayal of the sensitivity of each of the uncertainties on the portfolio metrics.

Infrastructure, technical, and people capabilities

Project-level data that is reliable and trusted by the organization is the foundation for any project or portfolio-level analysis. A consistent framework for project-level assessment and the associated data repository and data integrity system is essential to ensure a well-informed project-level decision in the context of the portfolio.

The communication of the insights to the leadership for decision making both at the project- and portfolio-level analysis should be in sync. A visualization dashboard with agreed-upon familiar views should be used to facilitate the process.

Soft and hard skills of the group running the portfolio management process ultimately will determine the level of adoption and success of the portfolio management in any organization.

<hr>

Table 3.1 Types of Uncertainties

<hr>

Local/project level

External/strategic

Competitors: Existing and new competitors

Substitutes: Unmet medical need; technological and treatment advances

Customers: Changes in buying patterns, user practices, payer perspectives

Suppliers: Availability of raw materials

Technology: Required for development and market success of new scientific
  breakthrough or discovery

Partners: Dependency on third parties

Regulatory: Changes in requirements

Social trends: Unmet needs and environmental concerns

Financial: Achievability of financial targets

Portfolio fit: Fit with portfolio strategy

Operational/process

Resources: Availability (quantity and quality) when needed

Time: Ability to enroll patients/meet timelines required

Team: Turnover of key team members

Budget: Exceed budgeted cost due to high level of risk mitigation plan costs
  more than the impact of the risk

Decision criteria: Clarity of go/no-go criteria

Information: Ability to gain access to needed information

Technical/scientific

Technical: Ability to meet safety, pharmacology, pharmacokinetic, chemistry,
  manufacturing, and controls (CMC) targets in the product profile

Regional/therapeutic/business unit level

Project- and macrolevel uncertainties that impact almost all projects within a
  therapeutic area or a business unit.

Global/portfolio level

Macro scientific, economic, political, and regulatory phenomena that impact
  almost all projects in the portfolio.

<hr>

The portfolio management is an evolutionary process and requires an ongoing balance between the organization's culture readiness and installation of sophisticated tools and methodology for portfolio management rigor. Patience and a mindset for dynamic evolution and a learning loop are essential.

A mature portfolio management organization has a central data source for all project- and portfolio-level data that are merged and linked for a real-time dynamic scenario planning to proactively understand and manage the portfolio uncertainties.

Best practices in portfolio management

Annual portfolio prioritization versus dynamic portfolio prioritization

Typically, the portfolio prioritization process follows the annual budgeting process. Once a budget is set for the following year, the projects in the portfolio are prioritized for funding, keeping in mind the budget constraint for the following year. Throughout the year, projects are approved on a first-come, first-serve basis for funding solely on their individual merit, that is, in isolation to portfolio.

Best practice: Portfolio management should not be an annual process; rather any project decision should be made in the context of the portfolio impact. Some companies are shifting from just the annual to a dynamic process for more efficient management of portfolio investment. Every project that is brought forward for funding approval is assessed individually with varying strategic themes to ensure it meets the minimal threshold and it is also assessed for portfolio fit. The project assessment as well as the portfolio fit assessment is compared with (1) currently unfunded projects and (2) projects expected to come for funding in the next 6 to 12 months. Based on the ability, when compared to currently unfunded and potential future projects, of the project to enable the portfolio goals, the project is either funded for the project option that provides the right portfolio fit or terminated/out licensed or put on hold to create an option that might be exercised later.

A similar exercise is conducted to reassess portfolio health and value when a project unexpectedly exits the portfolio.

Addressing immediate constraints versus managing constraints for the longer term

Typically, during the annual portfolio prioritization, projects are funded in order of their "value" until the budget for the next year runs out. Limited visibility is provided on the portfolio resource demand post the following year.

Best practice: Some companies have realized that taking portfolio-level decisions to manage just the following year's budget is an inefficient approach and in most cases this approach selectively funds projects with lower next-year funding but with higher total funding needed for future years. Thus every year due to the limited funds, the projects below the funding cutoff line are either terminated or left unfunded resulting in value loss. Accounting for 3 to 5 years of the funding profile in the portfolio decision ensures that the portfolio is managed to the real constraints. Thus, any consequences of including projects with higher later-year funding needs are clearly understood and plans could be put in place to proactively address the budget issue.

Balance between the qualitative and quantitative approach

Drug development is a highly uncertain process with a decreasing level of uncertainty from the discovery to the registration phase. The project- and portfolio-level decision process should be respectful of this and use a right balance of qualitative and quantitative analysis to rightly inform portfolio decisions.

The projects in the discovery and early stage have high uncertainty and a low level of visibility into the inputs that go into a classical financial valuation. Hence, the projects in the discovery and early stage will be more suited for a qualitative assessment for the attributes (unmet need, innovation, strategic fit, level of investment, time and cost, etc.) that are critical to meeting portfolio and company strategic goals.

Projects about to reach PoC (proof of concept) and in post-PoC stage are more suited for a quantitative financial evaluation. A point of caution is that project- and portfolio-level decisions should not be made merely on the financial merit. Financial approach (like efficient frontier) is a good starting point to understand the order of buy-in based on the highest value under resource constraints. The financial metrics should be looked at in conjunction with other qualitative attributes and interdependencies to truly model the portfolio risk and settle on a portfolio that provides the best chance of meeting the company strategy.

Including projects from discovery to lifecycle (end to end) for portfolio prioritization and optimization also ensures that any interdependencies (e.g., lead/backup compounds, projects targeting same patient/indication)

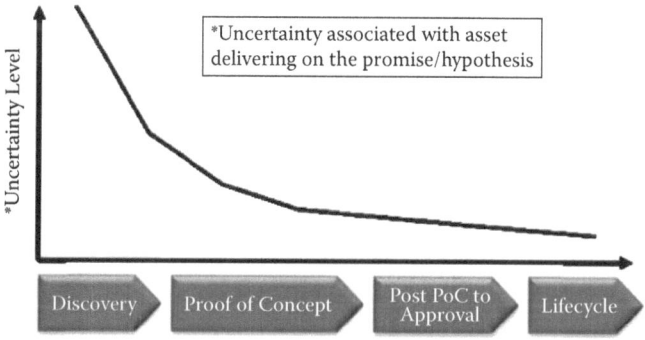

Figure 3.3 Level of uncertainty reduces as projects progress toward approval.

and associated technical and commercial risks (e.g., "self-cannibalization") are well characterized in the ability of the portfolio to meet company goals.

Illustrative framework for approach and decision level for portfolio models

Assessment of project-level risk

Scientific and medical risks contribute the most to the overall uncertainty of a pharmaceutical portfolio. Since these risks are conceived at the project level, a consistent methodology to quantify these project-level risks, in a transparent fashion, is essential to assess their impact both at project and portfolio level.

Best practice: Some companies have realized the importance of this and have implemented processes that provide the needed visibility into key sources of technical risk and its connectivity for portfolio-level analysis. The key steps in project-level risk assessment include:

- Relevant "risk categories" and associated "risk factors" are identified and agreed to by the leadership for both project- and portfolio-level analysis.
- As a project progresses, the subject-matter experts on the project make quantitative assessment of each risk factor and the risk category.
- The project lead then assigns an overall risk score to the project taking into account the assessments made by subject-matter experts.
- The assessment scores at risk factor, risk category, and project level are available for project- and portfolio-level analysis.

Illustrative Framework for Approach and Decision Level for Portfolio Models

Sub-Portfolios	New Candidates	Early Stage Programs			Late Stage Programs		Extension (LCM)
	Discovery	PC	Ph I	Ph IIa	Ph IIb	Ph III	Ph IV
% Drop Out (Benchmark)	>95%	>95%	90%	80%	60%	50%	25–50%
Level of Decisions	DA choice (given) and competitive strategies	Program priority and indication selection			Product profile and strategies		Line extensions and enhancement
Decision Units	DA (given), MoA and Targets	Program/Compound/Indication			Product Profile at Launch		Product claims post launch
Parameters	Market potential, share, barrier of entry	Time to PoC/market PTRS Lead/backup		Indications, line of therapies	Profile (indications, efficacy, safety and presentation), clinical, regulatory and commercial		Incremental revenue and costs
Portfolio Objectives	Strategic intent and goals	Shots on goals, Speed to PoC, Improve POS			Financial risk and return, time vs value		
Time Horizon	~30 yrs	~20 yrs			10–15 yrs		5–10 yrs
Asset Assessment	Aggregate	Aggregate/Assumption - Based			Project Specific		

Figure 3.4 Illustrative framework for approach and decision level for portfolio models.

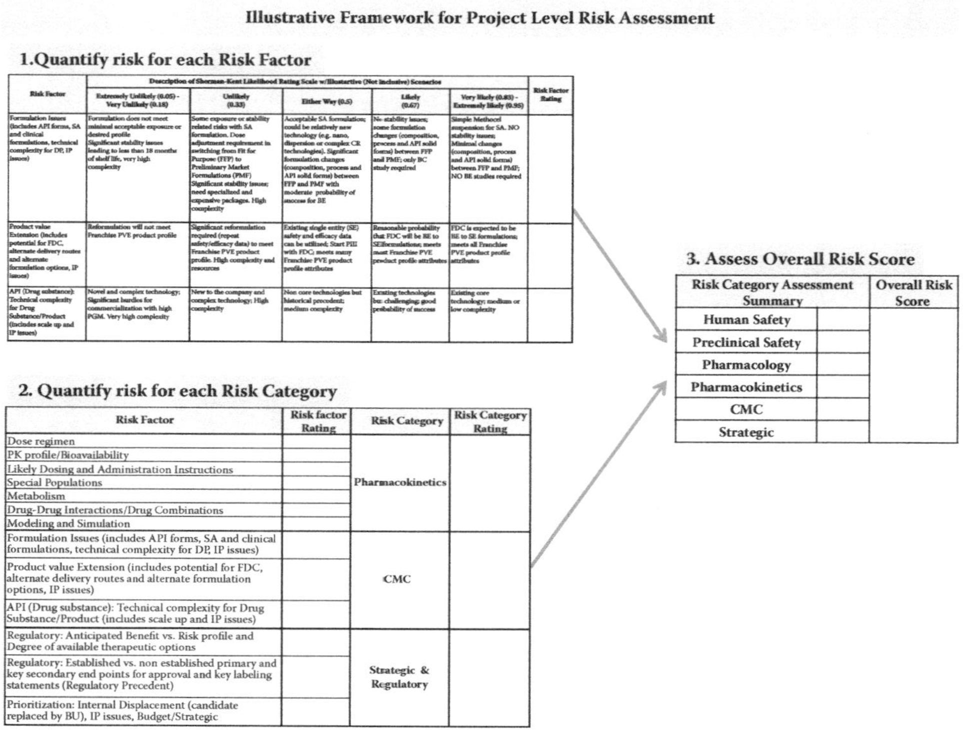

Figure 3.5 Illustrative framework for project level risk assessment. *Source:* An "Expert System" Approach for Uncertainty Assessment Enables Transparency and Consistency for Project and Pipeline Level Prioritization, Arun Kejariwal presentation at CBI Strategic Pipeline Planning, September 28–29, 2010, Princeton, NJ.

Scenario or what-if analyses should be integral to portfolio management.

It is virtually impossible to predict the future so it is important to develop a portfolio strategy that is robust enough to overcome the shock of most unplanned events. Decisions based on insights from discrete scenario planning both at the project and portfolio level increases the chances of the portfolio meeting its goal.

Projects should be assessed for multiple options accounting for decisions that are in the team's control. Each project option should also be discretely modeled and pressure tested under different future scenarios based on the potential outcome of key micro- and macrolevel uncertainty. Different options and the modeled outputs along with commentary on the weak and strong points should feed into assessing the "portfolio fit" for the project.

Similarly, different portfolio options should be discretely assessed for different future environments and then a portfolio that can withstand most potential scenarios and also meet the strategy objective should be picked.

This type of scenario analysis should account for key uncertainties that the organization has identified and has reached alignment. The portfolio implications, both upside and downside, of the identified uncertainties and any uncertainties not explicitly considered should be clearly communicated.

Questions should be asked about what could be done differently with projects in the current portfolio to better withstand future scenarios. Should the company stay the course, expand scope, reduce scope, partner, and so on?

The pitfall of the analytics behind portfolio simulation is that it narrows the organization focus to the most probable outcomes. In contrast, the discrete scenario planning ensures that the organization is able to proactively identify signposts and plans to monitor and adjust the portfolio mix as the dominating future scenario evolves, thus providing organization a dress rehearsal and further instilling nimbleness into portfolio management.

Conclusion

A mature dynamic portfolio management system requires an integrated project and portfolio system with committed organizational leadership, credible qualitative and quantitative valuation methodology, range of investment alternatives, credible inputs, excellent communication and collaboration, sound process, effective process leadership, and appropriate software systems.

A robust portfolio management system should be able to answer the following key questions:

- How does each opportunity in the portfolio contribute to the portfolio goal?
- Is the portfolio balanced with regard to our priorities?
- Will the portfolio support the near- and long-term growth targets and enable company strategy? Can we do it with the available resources over that time horizon?
- Are the trade-offs acceptable? Is the portfolio risk balanced?
- Is the overall risk of missing aspirations tolerable?
- What are the key uncertainties that need to be monitored and when do they get resolved?

Parting thought: Patience and discipline is necessary to reap the benefits of the systemic portfolio management approach. It is easy to lose patience at the instance of first failure/fear and quickly slip back into the easy path of ad hoc decision making.

Acknowledgments

The insights presented in this chapter include the learnings on portfolio management that I have gained through my interactions with colleagues and friends over the years. I would like to acknowledge their indirect contribution to this chapter. I apologize in advance for likely missing a few people who belong here: Lifei Chang, Suzanne Brown, Richard Bayney, Steve Chamberline, Christine Cioffe, Prasanna Deshpande, Jeffery Handen, Mathew Hendricks, Jack Kloeber, Andrew Levitch, Jonathan Mauer, David Robinson, Julie Schiffman, Nakin Sriobchoey, Rodger Thompson, and Daniel Zweidler.

References

Bayney, R., and Chakravarti, R. (2012). *Enterprise Project Portfolio Management: Building Competencies for R&D and IT Investment Success*. Plantation, FL: J. Ross Publishing.
Salo, A., Keissler, J., and Morton, A. (eds.). (2011). *Portfolio Decision Analysis: Improved Methods for Resource Allocation*. New York: Springer.
Sanwal, A. (2007). *Optimizing Corporate Portfolio Management: Aligning Investment Proposals with Organizational Strategy*. Hoboken, NJ: John Wiley & Sons.

chapter four

Funding and resourcing clinical development

Michael A. Martorelli

Contents

Introduction

In the dynamic world of clinical research, any written work spotlighting the latest developments involved in funding and resourcing the trials of potential new medications is likely to be a bit outdated on the day it is published. Indeed, there were several significant news items that should affect the nature of clinical development between the time this chapter was started and the date it was submitted for publication. Nonetheless, this chapter presents some thoughts on several areas in which tomorrow's clinical trials are likely to be different from yesterday's. At the outset, it is important to note the recent (and ongoing) initiation of several long-term strategic partnerships between drug sponsors and contract research organizations (CROs). These arrangements are fundamentally changing the relationship between those two parties, since many of them call for the sponsor to transfer unprecedented amounts of operating and resourcing risk to their CRO partner(s). In turn, those CROs are assuming more autonomy and responsibility for conducting trials. In light of that new answerability, they are beginning to play a more proactive role in transforming clinical research. Several implications of this evolving relationship between sponsors and service providers are noted throughout this chapter. Perhaps the most important is the changing motivation of both parties to address some historic bottlenecks in clinical research, thereby making the process both more efficient (very possible) and more effective (more challenging).

Funding clinical research

Emerging types of organizations are supplementing traditional groups of financiers in providing funds to support clinical research. Drug and biotechnology companies have long provided most of the funding that supports the discovery and development of promising new pharmaceuticals. Organizations such as the Pharmaceutical Research and Manufacturers of America (PhRMA), the Tufts University Center for the Study of Drug Development (CSDD), and Burrill and Co. continue to track annual expenditures by such firms of about $70 billion. The manufacturers of medical devices and certain diagnostic instruments spend smaller amounts developing their products, most of which require less extensive clinical research than drugs. Most sources suggest that about two-thirds of the drug industry's spending on research and development (R&D) supports the funding of clinical trials for potential new products. The National Institutes of Health (NIH) also funds clinical research in the amount of almost $30 billion annually. Both the drug industry and the NIH have been restraining the growth rate of their R&D spending (and sometimes even decreasing it) during the past several years, as budgetary pressures have caused their managements to examine more closely the size and nature of those budgets.

The data on the funding of clinical research shows some rather striking increases in R&D spending by other organizations. According to Research!America, in 2011 universities, independent research institutes, philanthropic foundations, and voluntary health associations spent more than $15.2 billion on R&D; back in 2002 the total for such organizations was only $8.2 billion. While those dollars still represent only a fraction of total spending, in certain therapeutic categories the sponsoring organizations have become particularly important participants in the R&D enterprise. The Michael J. Fox Foundation, Melanoma Research Alliance, and Myelin Repair Foundation are just a few groups that have helped spur previously underfunded research into specific disease states.

It is not yet clear what type of ownership rights and marketing royalties these organizations will be demanding from licensing or other codevelopment partners. But it seems they will become even more important participants in funding clinical research in their chosen therapeutic categories. Indeed, they may be filling a critical void. In reengineering their own product development efforts, many large drug companies are continuing to reorient their research efforts to focus only on areas with the largest economic potential. Thus, it seems likely those traditional product development firms will be spending less time, money, and effort conducting clinical research on drugs for orphan indications and diseases of the developing world. Patients with those conditions will need more help than ever from patient advocacy groups, charitable foundations, and independent research institutes. It remains to be seen how much if any slack in the rate of R&D spending by traditional drug development organizations can be offset by increases in spending, especially on clinical research, by such groups.

Individual drug companies have been forging codevelopment or comarketing partnerships for at least two decades. Over the years, many studies conducted by consultants, market researchers, and government bodies have suggested one tangible way to improve the productivity of the drug development enterprise would be to encourage more collaboration among its participants. Thus, companies and universities that considered all their research proprietary a decade ago have recently begun forming various types of collaborations and partnerships. Many collaborative efforts in the preclinical research arena have contributed to the progression of basic research on specific diseases or biological pathways. Few have resulted in the more efficient and rapid clinical development of a specific therapeutic product. Recently, however, many large drug companies have greatly expanded their relationships with academic institutions, biotechnology partners, and even other large drug company competitors. Firms long known for their condescending attitude to research conducted outside their corporate walls have created internal partnership programs that promise to blend the knowledge possessed by a range of collaborators with their own internal research.

Several new types of multipartner collaborations have the potential to be more effective in helping their participants leverage their own capabilities with those of others, and improving the team's success rate of discovery and clinical development. One particularly intriguing partnership involves the recently formed TransCelerate BioPharma, a collaboration of ten drug companies. It has selected clinical trial execution as its initial area of focus. Its members are working on several initiatives, including the development of an industry-wide approach to risk-based monitoring, and a shared, cross-industry investigator portal. Groups such as the Innovative Medicines Initiative, TB Drug Accelerator, and Medicines for Malaria Venture are other organizations jointly funded by multiple parties from the public and private sectors. It is not yet clear whether traditional drug companies participating in these ventures will be limiting their own spending on R&D as they increase their commitments to such partnerships. As is the case with patient advocacy organization, the parties have not yet been forced to look closely at the potential profit split of any successful development efforts initiated and largely funded by a multiparty organization.

Quintiles Transnational and PPD Inc. are two CROs that have experimented with funding clinical research programs for investigational drugs whose rights they had acquired. In the late 1990s, Quintiles formed a separate subsidiary whose management purposely sought new product candidates with strong commercial potential and that had successfully passed their initial screening trials. At that time, PPD also allocated a limited amount of personnel and financial resources to finding such products. Both firms acquired the development and marketing rights to several drugs, and used their internal resources to conduct the requisite clinical trial programs. Both managements made the case to investors that the potential financial returns of a successfully marketed drug exceeded the returns available to a company that limited itself to providing drug development services to others. Investors responded negatively, asserting that they believed services providers should neither seek nor accept the "molecule risk" of the products they were helping clients shepherd through the clinical trials process. That reaction led the management of Quintiles to take the firm private in 2003. In 2006, the company sold the remainder of its royalty rights for the antidepressant Cymbalta to the private equity firm TPG-Axon Capital Management. In 2002, Quintiles had struck a deal with Eli Lilly to co-promote that product in the United States. In 2009, in another move to distance itself from direct involvement in drug development, the company spun off to shareholders the PharmaBio subsidiary that owned and managed the company's investments in risk-sharing arrangements. Negative investor reaction was also behind the decision by the management of PPD to divest the company's compound development subsidiary. In 2010, it spun off that unit, renamed Furiex Pharmaceuticals, to shareholders. The point is not to critique the appropriateness of investors' views but to note the historic willingness of at least two service providers to engage in substantial risk-sharing for selected compounds. The question is being asked anew: What role should a service provider play not only in helping a sponsor conduct clinical research but also in accepting part of the "molecule risk" of the sponsor by taking an equity ownership position in the compound? Doing so would put that provider in the position of placing its own funds at risk in the conduct of clinical research, a debatable proposition.

Insurance companies of various types have long been paying for the majority of Americans' prescription drugs. Some large insurers and managed care organizations entered the clinical research business through acquisition or strategic partnership several years ago. However, those organizations did not mine their patient databases for potential clinical trial participants. Neither did they connect the dots and use evidence from clinical research to drive decisions on their reimbursement policies for drugs in categories with many alternatives. A 2009 Deloitte report titled "What Payers Want: Viewing Payers as Customers" outlined the need for such evidence. It cited the experience of the Regence Group, operator of Blue Cross/Blue Shield plans in the Pacific Northwest, in conducting its own studies to ascertain the usefulness of certain medicines. Meeting and publications sponsored by the International Society for Pharmacoeconomics and Outcomes Research (ISPOR) have become more important than ever. In this modern world of cost restraints, new types of combined provider–payer organizations are beginning to demand such evidence, and even conduct the appropriate trials, as they attempt to rationalize their high level of spending on pharmaceuticals. They are enhancing their capabilities to conduct more phase IV research within their own organizations and financing additional trials in the community. In contrast to traditional clinical trials that typically only evaluate a product's safety and efficacy when compared to a placebo, these provider–payer-directed trials frequently compare one newly approved product to existing forms of therapy. Armed with such comparative effectiveness results, those organizations hope to be able to adjust their pharmaceutical usage and reimbursement policies to encourage only the use of the most effective treatments for various conditions. Thus, they are becoming more interested than before in closing the loop and using clinical research to guide decisions on drug usage in a way that has rarely been done in this country. A May 2012 PWC Health Research Institute report, "Unleashing Value: The Changing Payment Landscape for the US Pharmaceutical Industry," notes that its recent survey found that 16% of payers currently have an outcomes-based contracting arrangement with pharmaceutical companies and one-third expect to support them within three years.

Such organizations may be helped in that effort if drug companies begin to disclose more information from the clinical trials of many existing medications. In January 2013, an unusual group of organizations collaborated on the creation of a website at www.alltrials.net and began urging all drug companies to upload all the data from all their trials of currently marketed products. They believe more transparency of clinical trial results will help physicians, pharmacists, payers, and patients make more informed decisions about the most effective product to use in treating various conditions.

In closing this discussion of funding clinical research, we address some changes in the *modus operandi* of many venture capital and private equity firms. Members of that community have long been important supporters of emerging drug companies, whether they were developing a small molecule or a drug based on biotechnology. During the past few years, data sources such as The National Venture Capital Association, PricewaterhouseCoopers, Thomson Reuters, Venture Source, Burrill and Company, BioWorld, and Silicon Valley Bank have chronicled the quarter-to-quarter variations in dollars invested and companies funded, and provided information broken down by stage of development, by specialty, and by stage of financing. Interpreting the inconsistently defined data can be challenging. The evidence can be read to suggest that the level of venture investing in drug development has either declined or remained consistent, that it is either easier to obtain funding for a start-up or that most investment dollars are going to already established companies, and that there are fewer investors funding new drug development than several years ago.

Regardless of the set of numbers one chooses to believe, it seems obvious that several trends are affecting the funding of emerging drug development companies. Traditionally, venture capital and private equity firms have provided their investment dollars to such companies without any specific constraints. The fund managers have taken only a moderately active oversight role, preferring to give most managerial authority to the companies' executives. Recently, however, venture financiers have been taking a more active role in helping select compounds around which to form companies. Representatives of private equity firms investing in emerging and established companies alike have also been exerting more influence over the nature of those companies' clinical research programs. Thus, in a growing number of emerging drug companies, financiers are directing the nature of clinical research programs; their companies might not be pursuing the compounds that are most preferred for development by the clinical research professionals on staff. The long-term consequences of this trend are unclear. Perhaps tomorrow's researchers will be too focused on targets with meaningful economic potential and not focused enough on products with less commercial appeal but more apparent biological or medical utility. With funding at a premium, however, and harder to obtain than heretofore, it is apparent that the providers of capital for clinical development will be exerting more influence over the nature of the industry's clinical research efforts.

High costs of recruiting clinical trial patients

Companies pursuing clinical research need to take new measures to reduce the costs of recruiting clinical trial patients. Study after study demonstrate the weakness of the historic procedures used by sponsors and service providers alike to recruit and retain appropriate patients for clinical research. Unfortunately, despite the importance of the topic and the acknowledged need to improve the speed, enhance the quality, and lower the costs of recruiting and retention, it can be difficult to nail down the numbers that would quantify the problem. For instance, although many sources place the cost of clinical research at approximately two-thirds of all pharmaceutical R&D, not many have rigorously studied the cost of retention and recruitment as a component of those costs. One provocative 2012 report by the consulting firm RDP Clinical Consulting suggests those costs may account for as much as 40% of the total budget for most trials. In the absence of multiple rigorous studies of that issue, it seems unlikely that the typical study's beginning budget would reach that level. Yet, it seems conceivable that the admittedly high costs of rescue activities in studies with only partially completed enrollment goals may indeed contribute to that seemingly exaggerated cost of the total recruitment activity. Of course, the point is not to document how much such efforts can cost in the most extreme cases but to illustrate the lack of data that would accurately quantify the true costs of the widely acknowledged problems of recruiting.

Academic groups such as the Tufts CSDD, publications such as *Applied Clinical Trials* and *CenterWatch*, consulting firms such as Cutting Edge Information, and patient advocacy groups such as CISCRP (Center for Information and Study on Clinical Research) have all investigated this topic and have all suggested that improving the speed and accuracy of recruiting patients can greatly improve the efficiency of clinical research. During the past several years, multiple groups have cited a range of informative statistics that suggest the dimension of the problem. These "facts" should each be considered best estimates on average over time rather than precise up-to-date quantifiers of the various items they address.

- About 90% of studies meet their enrollment goals.
  - But the average dropout rate is around 30%, and
  - 85% of studies fail to retain enough patients, so
  - nearly 80% of them fail to finish on time.
- More than two-thirds of sites fail to meet their enrollment goals.
- Up to 50% of sites enroll only one patient or no patients at all.

Another set of "facts" provides some rational justifications for the industry's difficulties in recruiting a sufficient number of appropriate patients for its clinical trials:

- Only 6% of the population will ever participate in a trial.
- About 94% of the populace has never been informed about their suitability for a trial.
- Less than 4% of physicians actively participate in conducting trials.
- More than 60% of eligible patients who refuse to enroll in a trial cite the risk of randomization (i.e., getting a placebo) as the major reason.
- Less than 10% of the responders to the typical advertisements about a trial become a randomized participant in that trial.

One last "fact" gives voice to the most recent evolution in the ways companies recruit patients for clinical trials.

- Only 15% of studies incorporate nontraditional approaches to recruiting.

Conferences and publications are making more references to the use of social media websites such as Facebook, MySpace, and Twitter, as well as online patient communities such as PatientsLikeMe, Inspire, and WeAre.US as potential vehicles for recruiting patients in clinical trials. The absence of formal guidance from the Food and Drug Administration (FDA) about using such sites for that purpose may be affecting the speed at which the pharmaceutical industry will be adding these tools to its recruiting toolbox. However, it is noteworthy that institutions such as the Mayo Clinic, Quorum Review IRB, and Pfizer have been outspoken in their belief that using these avenues to attract more patients to appropriate clinical research studies is quite consistent with the FDA's regulation of that effort.

Two other routes for improving the industry's recruitment problems relate to elements of clinical research funding noted earlier in this chapter. The first involves patient advocacy groups that are becoming more active in financing clinical research; they also need to become more active in helping recruit patients to participate in those studies. Many organizations possess registries of patients afflicted with their disease of interest and support clinical trial matching services for targeted therapies. Most forge close relationships with groups of specialty physicians treating patients and work in the community to raise the awareness of a particular disease or condition. They also advocate for more government attention to basic, translational, and clinical research in their chosen area of focus. These groups certainly have a role to play in encouraging more patients to participate in clinical trials.

Nearly one-third of the drugs approved in 2012 were for cancer indications. According to PhRMA, about 40% of existing trials involve potential cancer treatments, many of which are personalized targeted therapies. Medical oncologists are more actively involved in clinical research than physicians in most other areas. Yet, the American Cancer Society estimates that only 2% to 4% of today's adult cancer patients participate in a clinical trial. Doubling that percentage over the intermediate term should be the goal of every cancer treatment, research, and advocacy group in the country.

While discussing the expanded role patient advocacy groups could play in recruiting patients for clinical trials, it is also worth noting the "breakthrough therapy" provision of Prescription Drug User Fee Act (PDUFA) V, signed into law in July 2012. That clause permits the secretary of Health and Human Services to expedite the development and review of a drug intended to treat a life-threatening disease. The secretary is to issue draft guidance documenting the requirements and procedures involved in reviewing breakthrough products by February 2014. Groups such as the Abigail Alliance and backers in Congress have long sought expanded access to investigational drugs through existing compassionate use, expanded access, and managed access programs. The new law, possibly in combination with a proposed new regulatory pathway described later in this chapter, could make it easier for more patients to receive important drugs before their formal FDA approval.

The second route for improving the industry's recruitment problems requires insurance companies, managed care organizations, hospitals owning physician groups, and accountable care organizations to take a more active role in helping recruit patients for clinical research. Historically, even such firms that have entered the CRO business have been neither interested nor successful in tapping their patient populations for potential clinical trial participants. Many executives have noted the lack of financial incentive to attempt to match patients with trials; others have noted their organization's lack of infrastructure to pursue this objective; still others have suggested that privacy issues and family desires are important impediments to this endeavor. Ironically, it is possible that the more widespread disclosure of all the results of all clinical trials may make it even more difficult to recruit patients. It is conceivable that the widespread dissemination of such data will spotlight the large number of clinical trials that fail to reach positive conclusions or support the further development of a specific molecule. Regardless, given the industry-wide imperative of finding more appropriate clinical trial patients, the leaders of the aforementioned treatment and payer groups must find ways to become more involved in the clinical research enterprise.

An interesting June 2012 survey by McKesson and the patient advocacy organization CISCRP suggested that pharmacists may have a previously unrealized role to play in clinical research. More than 97% of the consumer respondents said they had never asked their pharmacist about a clinical trial, although 80% said they would want those trusted medical professionals to tell them about any relevant study. Although 87% of the pharmacist responders believe it is important to educate their customers about clinical trial participation, only 56% would be "very willing" to provide such information if those customers were interested in receiving it. Based on the survey, McKesson conducted a pilot study to look more closely at the role of community pharmacists in proactively seeking to connect customers with appropriate clinical trials. It mapped its 2,300 pharmacies to nearby investigative sites and identified potential clinical trial candidates for them from prescription claims data. It asked the pharmacists to vet the potential candidates and to contact the appropriate ones about participating in a trial. The 300 participating pharmacists referred 221 patients to a range of phase III and phase IV studies; 161 (73%) of those patients enrolled in a study. Perhaps a broader effort by companies such as McKesson, retail pharmacy chains, and appropriate trade organizations could work with drug sponsors to provide the right combination of education and incentives to help more community pharmacists add their respected voices and talents to the clinical research enterprise.

Any discussion of ideas to improve the recruitment of clinical trial patients would be incomplete without noting the growing importance of social media techniques. Pfizer's highly publicized mid-2011 experiment with social media to recruit patients from home demonstrated the challenges of such an effort. Online communities such as Inspire and PatientsLikeMe, and notices on sites such as Craigslist, Google Search, and Facebook generated substantial traffic on the website for Pfizer's Phase IV trial of the overactive bladder drug Detrol. However, the conversion rate from interested persons to clinical trial participants was very low. Conventional recruitment efforts involve guidance and information from a physician, investigative site, or patient advocate; prospective patients who were not able to receive such guidance proved unwilling to enter the trial without it. Hindsight also challenges the concept of finding patients for the trial of a condition with many relatively effective treatment options, and for the testing of social media recruiting for a condition most common to older women who might not be very technically savvy. Yet, it would not be prudent to dismiss the concept of social media recruiting. Research shows many people use the Internet to investigate various health and disease conditions. Sponsors must continue exploring creative new ways to tap into those experiences and drive more persons to the world of clinical trials.

Complexity of clinical trials

Sponsors need to design clinical trials with a sharper eye to efficiency, effectiveness, and practical realities. Two solvable problems behind the high cost of clinical trials are the extraneous amount of data collected and the large number of protocol changes. The aforementioned Tufts CSDD has done groundbreaking work documenting the nature of both problems.

- On average, 20% to 25% of all clinical trial procedures are considered noncore, that is, they are not used in the typical new drug application (NDA), which is after all the rationale for conducting most clinical research. Tufts estimates the drug industry spends more than $1 million per trial or a total of $4 billion to $6 billion each year on procedures that result in the generation of extraneous clinical trial data.
- Protocol changes are often necessary elements of a clinical trial. But Tufts' data suggests that one-third of all amendments are avoidable. It places the cost of dealing with these unnecessary changes at approximately $2 billion per year.

Both problems relate to previously unacknowledged flaws in the traditional processes used to plan and conduct clinical research. The evidence shows that the average trial incorporates more procedures than ever before, and that trials for drugs to be used to treat chronic conditions involve more complex testing and analysis than those for drugs aimed at treating acute conditions. When evaluating a product that could potentially be taken for several decades, it may be necessary to spend more time evaluating its effects on all the body's biological systems, not just the one involved in the disease or condition. Moreover, changing regulatory requirements (or preferences) frequently cause sponsors to add procedures to ongoing trials in an effort to evaluate additional pieces of information not contemplated in the original protocol.

Historically, sponsors forced to design more complex trials to address specific medical and scientific questions have also produced prepared protocols that seem rich with academic imperatives but lack user-friendliness for the patients or the sites. Until very recently, most members of internal development teams worked in an insulated world of fellow drug development scientists and had little connection with physicians who conducted clinical trials or treated patients. Those teams have often been pressured by their companies' senior executives to drive their clinical development plans more forcefully and adhere to self-imposed deadlines. They have been forced to rush many complex protocols through rubber-stamp approval committees whose members may not have fully understood their lack of relevance to the clinical practice of medicine. As a result, as many as 40% of protocol amendments have been written after the trial has commenced but before the first patient has been enrolled; such changes frequently require changes in eligibility criteria and extend the time to first-patient-in by more than 100 days.

In the recent past, numerous articles in publications such as *CenterWatch* and *Applied Clinical Trials* have spotlighted the changes that many drug and biotechnology firms have been making to ensure that their protocols were developed with every consideration for medical, scientific, and operational feasibility. The recently formed Society for Clinical Research Sites (SCRS) also appears to have a role in proactively consulting with sponsors to help them develop protocols that can be efficiently and economically implemented by the sites participating in the study.

Executing trials more efficiently and effectively

Sponsors hoping to increase the efficiency and effectiveness of clinical research need to remain at the forefront of using a range of new tools and techniques. A shrinking number of clinical researchers remember the clinical trials of yesterday when a drug company would use internal personnel to manage every aspect of a study. From the mid-1970s to the mid-1980s, pioneers such as Hein Besselaar, Dennis Gillings, John Schrogie, and Josef von Rickenbach left the drug industry or academia to establish companies that would perform some of those services on a contractual basis, thus creating the CRO industry. In those days, reluctant drug company personnel used such outsourcing firms largely to cover peak periods of activity or to help manage work on low-priority projects that internal people considered nonessential dead ends. Throughout the 1990s, sponsors were facing an unprecedented set of financial challenges, brought on by discussions about national health insurance, the prospect of important patent expirations, and the reality of lagging productivity in their R&D efforts. As they increased their use of outsourcing, most companies discovered two important financial benefits of this action:

1. CROs provided staff and technical expertise only on an as-needed basis; a drug company could reduce the fixed costs of its internal staff members and replace them with the variable costs associated with the use of an outsourcing firm.
2. Upon the completion or termination of a project managed by internal resources, a sponsor would have to carry those employees until they could again be productively employed; the staff of a CRO managing a completed or terminated project could be terminated very quickly.

Throughout the late 1990s and early 2000s, clinical development specialists working for many drug companies still seemed to view outsourcing primarily as a cost management tool that was being forced on them by financially oriented corporate executives. Too many researchers in that industry regarded outsourcing professionals in CROs as "second-class" personnel; and too many CROs unintentionally confirmed that description by using newly minted coordinators and project managers, and tolerating an unusually high degree of employee turnover. Sponsors believed they needed to exercise an unusually tight degree of oversight, and in doing so, they usually failed to realize the full benefits that outsourcing should have provided. Many failed to recognize the inherent difficulties in using very limited pilot projects to ascertain the true costs and benefits of outsourcing. They also had to learn by trial and error exactly how much internal management oversight was required to oversee even those outsourcing projects that were being carried out efficiently and effectively by their CRO partners.

Attitudes toward outsourcing seemed to change for the better in the early years of the 21st century. Layoffs and restructurings within the financially challenged drug industry resulted in many experienced clinical development people moving into the ranks of the CROs. In seriously reconsidering the nature of their internal R&D efforts, many sponsors realized they could indeed use outsourcing firms to accomplish most of the tasks involved in shepherding a developmental compound through the phases of its clinical research program. Firms that had previously recognized some of the value in outsourcing began selecting several to become their "preferred providers" of various clinical development services. By 2005, many drug companies realized they could derive even more benefits from outsourcing if they would use a limited number of CROs to handle the majority of the outsourcing work they commissioned across the clinical research enterprise. Contracting in this way actually became more burdensome to many sponsors, however. They found that using a mixture of full-service and functional specialists still put a significant oversight burden on their own internal development teams.

As the decade ended, a few pioneering sponsors realized they could improve the efficiency of their clinical research programs even more by limiting the use of CROs to just one or two that would become their long-term strategic partners. A large share of today's clinical trials are being conducted by a relatively small group of strategic partner CROs; and those firms are finally using the latest electronic data capture technology in virtually all their new programs in order to enhance the efficiency of the voluminous "paperwork" requirements of the typical trial. Only a handful of the largest CROs with the broadest array of services have been capable of convincing sponsors to establish large-scale strategic partnerships. These arrangements are reshaping the CRO industry; the largest providers are taking business from midsized firms that previously had strong preferred-provider relationships with the largest sponsors. In order to stay competitive, most of those firms are maintaining an active acquisition search posture as they attempt to broaden their set of offerings or establish a presence in additional countries. A close reading of the industry's news flow would suggest that many of them have been achieving those objectives by hiring additional staff or opening new offices. Relatively small but highly specialized outsourcing firms do not appear to be as fundamentally challenged as their midsized brethren in continuing to generate new business from large and midsized drug development firms. Even the sponsors with only one or two strategic partners appear to have the ability to hire a small firm providing a high-quality specialty service whenever that seems appropriate.

If the drug industry is to maintain or increase its output of useful medicines, it must make tomorrow's clinical trials even more efficient than today's. Sponsors should pursue several different avenues to accomplish this objective: maximize the value well-qualified outsourcing firms can offer; take full advantage of the FDA's initiatives to improve the clinical research process; and pursue more "disruptive innovations."

Maximize value of outsourcing firms

The time has passed (if it ever was here) when drug company CEOs should be considering the conduct of clinical trials as their company's core capability. Data published by the Tufts CSDD documents the improvements that can be obtained by outsourcing most elements of the typical trial. Sponsors should not totally eliminate their clinical research departments. But they should insist that their internal program/project management personnel utilize to the greatest extent possible the preselected strategic partners to perform most trial management activities. They should definitely retain the right to seek alternate best-in-class service providers for selected critical functions. Optimizing the use of outsourcing providers requires sponsors to create an oversight structure that is effective but not overly dictatorial. Such firms' executive managers must be more willing to supplement the scientific and technical capabilities of their senior internal development people with an assortment of "soft" skills (e.g., communications, leadership, team building, mentoring, conflict management, flexibility in problem solving) not normally possessed by technically trained people who have spent their entire careers inside the drug industry. Many other industries employ people with practical experience in managing teams of vendors, partners, contractors, and internal personnel who are working on long-term projects. The drug industry should be more open to seeking out and hiring people with proven project management skills from other industries.

Take advantage of FDA initiatives

During the past few years, the FDA has shown a new willingness to help drug development firms navigate the regulatory pathway to a product approval. It has issued a range of Guidance Documents on subjects such as electronic submissions, enrichment strategies for clinical trials, the use of genomic information in early-stage clinical research, and the development of biosimilars. In February 2013, the agency held a conference describing its consideration of a new regulatory pathway to approve drugs for the unmet needs of patient subgroups with serious or life-threatening conditions. Limited trials could become the basis for extended "compassionate use" programs for products not intended for vast numbers of patients but targeted to small patient populations and conditions with no viable treatment options. The FDA is also more active than ever in working with organizations such as the Critical Path Institute, CDISC, and HL7 Clinical Operability Council to support the development of therapeutic area data standards. Agency topsiders are more visible than ever as attendees, exhibitors, and participants in industry conferences, as the FDA moves into the next phase of its 3-year-old Transparency Initiative. Drug industry veterans used to dealing with a regulatory body whose mission seemed to be preventing the approval of new products need to encourage newer members of their regulatory affairs teams to help the agency keep breaking down its communications walls.

Pursue more "disruptive innovations"

A brief list of potential new concepts should suffice to make this point:

1. Forego some intellectual property protections in an effort to be more collaborative in the basic research and discovery process.
2. Combine "proprietary" investigator databases.
3. Use "adaptive" trials all the time.
4. Use electronics to totally replace all clinical trial paperwork.
5. Move totally to risk-based monitoring.
6. Take aggressive steps to eliminate administrative inefficiencies that seem to underlie many clinical trial dropouts.
7. Insist that manage care organizations aggressively use electronic medical records (EMRs) to identify large groups of potential clinical trial patients.
8. Stop dealing with investigative sites that repeatedly fail to recruit sufficient numbers of clinical trial patients. (Ironically, one of the reasons drug companies insist on using those sites is because they are run by practitioners who are high prescribers. Hopefully, the weight of evidence-based studies, not the financial rewards of participating in a clinical trial, will become the most relevant tool to entice a practicing physician to use a medication over another.)

Of course, it is much easier to develop such a list than to implement any one of its elements. Perhaps the most important changes all participants in the clinical development enterprise need to make are (1) for managers at all levels to acknowledge the problems the industry needs to overcome, and (2) for all responsible leaders (and followers) to constantly strive for new ways to improve the *efficiency* of the clinical trials process. Achieving the more important but more elusive goal of improving the *effectiveness* of clinical research will require drug development firms to improve their selection of preclinical candidates to put through the extensive and expensive rigors of clinical trials. But that, as they say, is a subject to tackle on another day.

chapter five

Emerging role of patients in clinical research

Michael S. Katz

Contents

Integration of patients into clinical research is a work in progress, as the various stakeholders build relationships and learn how to leverage each other's strengths. In this chapter, we look at the evolution and impact of patient involvement in the cancer community, using case studies from the cancer community to illustrate best practices for integrating patients into the process.

As the ultimate end users, patients have always been involved in clinical research, if only as passive trial subjects or tissue donors. In recent years, patient involvement has become more active and substantive, encompassing the full continuum of clinical development, albeit

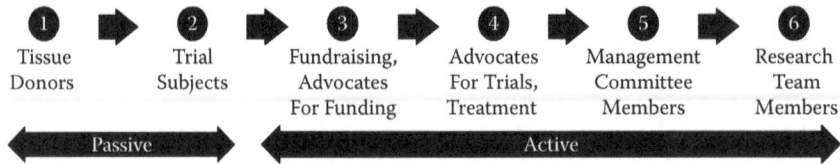

Figure 5.1 Increasing scope of patient integration into the research process.

only in limited segments of the clinical research community. The scope and impact of patient involvement varies widely across different research enterprises, both by the type of research, the disease or condition, and the funding source or sponsor. The scope of patient integration is increasing, fueled by demonstrable results and mandates by public sector funders (see Figure 5.1).

Caregivers often serve as surrogates for patients who are unable or unwilling to actively participate. Advocacy organizations generally involve patients but also bring others into the mix to work on behalf of patients. For brevity's sake, please read references to "patients" in this chapter to mean "patients, caregivers, and advocacy organizations."

The necessity of involving patients' as passive tissue donors or as subjects in clinical trials is well understood and accepted. There is no such consensus on involving patients as active participants in the planning and execution of clinical research. Full integration of patients as members of management committees and research teams is challenging, requiring education as well as changes in behavior and established practices, both for clinicians and scientists as well as patients. Patients typically lack the expertise and vocabulary to understand many of the issues confronting clinical researchers. However, in all fairness, clinicians and scientists typically lack the expertise, experience, and vocabulary to understand many of the issues confronting patients. When patients are integrated into research, both the patients and the scientists learn from each other. But, it does take time for this learning to occur. In a report to the National Institutes of Health (NIH) director, the NIH's Council of Public Representatives stated: "Community engagement is a core element of any research effort involving communities. It requires academic members to become part of the community and community members to become part of the research team, thereby creating a unique working and learning environment before, during, and after the research" (National Institutes of Health 2008).

Evolution of patient involvement

The bellwethers of patient involvement have been those dealing with serious chronic and incurable conditions (e.g., HIV/AIDS, diabetes, cancer, multiple sclerosis). These patients have become increasingly knowledgeable and proactive, having significant impacts via a combination of individual initiatives and through advocacy organizations.

Historically, patients played no role in clinical research and development (R&D) beyond participation as tissue donors or passive subjects in clinical trials. The HIV/AIDS community was the first to drive transparency and active involvement, born of anger and frustration at the lack of progress in combatting the AIDS pandemic. Cancer advocates, initially breast and prostate cancer advocates, followed suit, focusing first on public funding and subsequently getting into the substance of clinical trials design and conduct. As other cancer constituencies became more active and formed advocacy organizations, they became more involved in the research enterprise.

Advocacy organizations drove patient involvement by leveraging their substantial constituencies and professionalizing education and legislative initiatives. Government agencies, including the National Institutes of Health (NIH), National Cancer Institute (NCI), and the Food and Drug Administration (FDA), began including patient/consumer representatives on their advisory committees and review panels. The NCI, and more broadly the NIH, began building requirements for patient and consumer involvement into their cancer research grant programs, imposing requirements for patient involvement on their grantees. Other agencies and institutes soon followed, broadening involvement.

Although there are still substantial obstacles to overcome in broadening the scope and substance of patient involvement in clinical research, there is clear progress, especially in the public sector (see Table 5.1).

Private sector (e.g., pharmaceutical company) involvement of patients tends to be via sponsorship of publicly funded (e.g., cooperative group trials) or in the form of grants to advocacy organizations. Pharma also supports postapproval educational and market research events. Pharma does not yet routinely integrate patients and consumers into its preapproval research. There are substantial barriers, including confidentiality, that often block patient involvement in preapproval research. However, once research progresses to industry trials, the barriers are more about precedent, culture, and inertia than substantive obstacles. Of these obstacles, inertia is perhaps the most challenging, as there are long learning curves for patients about the science and for the entire research team on how to best leverage patient knowledge and capabilities. There is also no established mechanism for pooling experienced advocates across multiple pharmaceutical companies. The NCI has a central resource (i.e., the Consumer Advocates in Research and Related Activities program) for qualifying, training, and placing advocates in the many peer-review groups and committees across the institute. No such facility is available to pharmaceutical companies.

Table 5.1 Milestones for Consumer Involvement in Public Sector Research Activities

Agency (2012 budget)	Statements by agency about consumer/patient involvement	Patient roles
National Cancer Institute ($5 billion)	Consumers usually have firsthand experience as cancer survivors, or are relatives of cancer patients, or are active in cancer advocacy organizations. You have been selected on the basis of your involvement in the cancer experience; cancer advocacy experience; ability to communicate and advocate a position effectively.[a]	Director's Consumer Liaison Group (DCLG), and other federally chartered advisory committees, peer review, and management committees (e.g., Cancer Therapy Evaluation Program).
National Institutes of Health ($30.7 billion)	The Council of Public Representatives (COPR) is designed, based on Institute of Medicine recommendations, to (1) obtain the broadest public input to the NIH director on matters of public importance concerning biomedical research, research training, and the development and dissemination of science and health information to the public and (2) conduct the broadest outreach to increase the public's understanding of the NIH and its biomedical research programs.	Council of Public Representatives (COPR), a federally chartered advisory committee.
Food and Drug Administration ($2.5 billion)	The Food and Drug Administration (FDA) initiated the Cancer Drug Development Program to incorporate the perspective of patient advocates into the drug development process. This program provides patient advocates representing serious and life-threatening illnesses an opportunity to participate in the FDA drug review regulatory process. The Center for Drug Evaluation and Research (CDER), the Center for Biologics Evaluation and Research (CBER), and the Office of Special Health Issues (OSHI) in the Office of the Commissioner (OC) are participating in this program.[b]	Patient Consultant Program, review panels.

Table 5.1 (continued) Milestones for Consumer Involvement in Public Sector Research Activities

Agency (2012 budget)	Statements by agency about consumer/patient involvement	Patient roles
Coalition of Cancer Cooperative Groups	Each Coalition Cooperative Group member has a patient advocate committee to ensure that the patient perspective is integral to the design and implementation of Cooperative Group clinical trials, and to promote patient-centered advances in research and timely dissemination of research results.	Patient Advisory Board at the Coalition and similar groups at each of the publicly funded cooperative groups.
Department of Defense, Congressionally Directed Medical Research Program (CDMRP) ($378 million)	The unique voice and experiences of patients, survivors, family members and advocates play a pivotal role in the Congressionally Directed Medical Research Program (CDMRP). The innovative vision of research at the CDMRP integrates the experiences of consumers and the scientific community in the funding review process. Consumers are involved in all aspects of the review process. They add perspective, passion, and a sense of urgency that ensures the human dimension is incorporated in the program policy, investment strategy, and research focus.	Consumer representatives serve on all peer-review panels.

[a] National Cancer Institute, "The NCI Consumers' Guide to Peer Review," http://deainfo.nci.nih.gov/PeerReview/GuideCompleteBook.pdf, p. iii.

[b] Food and Drug Agency, "Drug Development Patient Consultant Program," http://www.fda.gov/ForConsumers/ByAudience/ForPatientAdvocates/PatientInvolvement/ucm123859.htm.

Why bother? What do patients have to offer?

There are those who think that patient integration is more about appearances than substance. Research is all about testing hypotheses, many of which have either a negative or lackluster result. The research enterprise is a hit-or-miss undertaking, as is patient integration. Like research scientists, patients can work years before having a major success. The impact of patient integration depends on the capabilities and behavior of everyone involved—and a little bit of luck.

Patients have real-world knowledge about their disease or condition and current treatment options. This knowledge stems from firsthand experience and discussions with other patients and clinicians. Patients are living with the disease or condition. They are singularly focused on timely development and approval of new, more effective treatments. As such, patients provide a perspective beyond the typical view of clinicians and scientists. This includes how current (and prospective) treatment options are viewed by the community—pros and cons, and strengths and weaknesses of the various options and modalities. Patients also have unique perspectives on side effects as well as dose modifications occurring at the grassroots level. Quality-of-life impacts and compliance issues as well as compliance issues are often better understood by patients. Scientists and clinicians generally view these issues at an aggregate level, through the lens of often-cumbersome and inconsistently executed survey instruments designed to get at these questions. Patients bring the grassroots perspective on current options as well as on unmet medical needs.

When patients are involved in the research process, either in management or on the research team, patient focus is dependent on the type of research and the state of the science in the targeted therapeutic area. Examples of the types of issues patients typically focus on are detailed in Table 5.2.

Patient advocacy organizations serve as arbiters of patient interests. There are omnibus organizations such as the American Cancer Society and Cancer Care. These organizations are most effective in dealing with broad policy and funding issues, funding research, and providing support services. More focused organizations get into the issues confronting patients battling specific diseases (e.g., breast cancer, prostate cancer) or conditions (e.g., autism). These organizations have therefore been the most involved in disease- or condition-focused research initiatives, including clinical trials.

Patients have driven dramatic improvements in drug development priorities and timelines, as well as increases in public- and private-sector funding for research. No database exists that would allow broad measurement of these patient-driven outcomes. Nonetheless, there are numerous examples of patient involvement having dramatic impact. To illustrate this, we present five case studies that illustrate how patient involvement can make a difference.

Case Study 1: Patient's Wife Reaches Beyond the Myeloma Community to Identify a Novel Agent (Thalidomide) That Ultimately Became the Standard Frontline Treatment for Myeloma

Case Study 2: Active Patient Involvement Drives Phase III E4A03 Trial Design That Changed the Global Standard of Care

Table 5.2 Patient Involvement in the Research Process

Potential application	Type of research		
	Basic science	Preclinical (e.g., in vitro)	Clinical trials
Epigenetics/ biology/ therapeutic targets	• What are the potential applications of this research for patient care? • How could the proposed research ultimately benefit patients?		
Prophylactic Palliative Improving outcomes Chronic control/ cure		• What are the potential applications of this research for patient care? • How could the proposed research ultimately benefit patients? • Does the treatment address an unmet need?	• How could the proposed research ultimately benefit patients? • Does the treatment address an unmet need? • How "accruable" is the trial? • What is the impact of symptoms, side effects, and monitoring on QOL? • How much time and travel will be necessary for patients to participate? • How do the protocol arm(s)' treatments compare with other options? • How credible are the hypotheses upon which the research is based? • Are there elements of the eligibility criteria that will be problematic for significant segments of the patient population? Is there good rationale for problematic criteria?

Case Study 3: Patient Leads Effort Using Consumer Focus Groups to Optimize Eligibility and Randomization Criteria for the TailoRx Breast Cancer Trial

Case Study 4: Patient Participation in FDA Oncology Drug Advisory Committee (ODAC) as Committee Members and as Speakers at New Drug Application (NDA) Meeting Influence Approval Decision

Case Study 5: Advocacy Organization Addresses Postmarketing Safety Issues with a Next-Generation Bisphosphonate

Case study 1: Patient's wife reaches beyond the myeloma community to identify a novel agent (thalidomide) that ultimately became the standard frontline treatment for myeloma

Multiple myeloma is a blood cancer, a malignancy of plasma cells. It is increasingly more treatable but remains incurable and, in most cases, fatal. Historically, myeloma patients could be treated successfully for an average of 3 to 5 years, after which the disease typically became refractory to treatment. With the disease out of control, kidney damage, bone lesions, or immunosuppression ultimately results in death. After decades of, at best, modest progress with various cytotoxic cocktails and bone marrow/stem cell transplants, myeloma research was energized by the arrival of novel agents, beginning with thalidomide in the late 1990s (Singhal et al. 1999). Thalidomide has a gruesome history, owing to its initial use in Europe almost 60 years ago as a treatment for pregnant women with morning sickness (Zimmer 2010). The drug caused horrendous birth defects because of its antiangiogenic effect (it prevented formation of new blood vessels, blocking normal development of embryos).

Dr. Ira Wolmer, a New York cardiologist and myeloma patient, reached the point where his disease was relapsed/refractory, with no viable treatment options. His wife, Beth, was not content to stand by and watch him die. She became a de facto member of the research team when she reached out to identify experimental treatments that might be an option for her spouse.

Mrs. Wolmer had read about Judah Folkman's research into cancer treatments using antiangiogenesis to prevent tumor growth. She approached Folkman and asked if one of the drugs he had been working on might help her husband. Folkman reviewed Dr. Wolmer's pathology slides and found that there was neovascularity in his bone marrow, confirming that antiangiogenesis therapy was a reasonable approach in his case ("Myeloma Today Profile" 2001). Folkman told Mrs. Wolmer that none of his drugs were ready for human trials. However, he did suggest thalidomide as a potential antiangiogenesis agent. Thalidomide had been approved in the United States in 1998 for the treatment of leprosy and was being marketed by Celgene.

Somehow, Mrs. Wolmer managed to convince the doctors at the Myeloma Institute for Research and Therapy at the University of Arkansas to seek permission from the FDA and Celgene to test thalidomide. Permission was granted for a limited trial with three relapsed/refractory patients, including Dr. Wolmer. Two of the three patients had dramatic responses to thalidomide. This was remarkable because the patients were heavily pretreated and refractory to all of the standard treatments. Ironically, Dr. Wolmer was the one nonresponder and subsequently succumbed to the disease. Based on the positive results with two of the three patients, a larger trial was initiated in which 84 previously treated patients with refractory disease showed a response rate of 32%, remarkable in refractory myeloma (Singhal et al. 1999).

Based on these results, thalidomide was approved for the treatment of multiple myeloma and broadly adopted. Patients were drawn to thalidomide because it was a new option for relapsed/refractory patients with no other viable options. Thalidomide was also attractive because it was an oral treatment. Further, it did not damage stem cells, which made it an ideal induction therapy for newly diagnosed patients planning to proceed to an autologous stem cell transplant. Success with thalidomide led to the development of two molecular analogs now approved for treatment of myeloma—lenalidomide/Revlimid™ and pomalidomide/Pomalyst™. Celgene's 2011 revenues for Revlimid were over $3 billion (Celgene Corporation 2012).

Pharma took notice of the commercial success of thalidomide and the accelerated development and approval process made possible by the unmet needs in relapsed/refractory myeloma. This created an impetus for commercial drug development investments in myeloma, leading to other treatments being tested and approved (e.g., bortezomib/Velcade™ and carfilzomib/Kyprolis™). Millennium, the developer of myeloma drug bortezomib/Velcade was acquired by Takeda Oncology. Proteolux, developer of myeloma drug carfilzomib/Kyprolis, was acquired by Onyx Pharmaceuticals.

Without the initiative taken by Mrs. Wolmer and her active role as an advocate for her husband and a de facto member of the research team, none of this would have been possible.

Case study 2: Active patient involvement drives phase III E4A03 trial design that changed the global standard of care

This is a personal story, as I, the author of this chapter, am the protagonist, a myeloma patient with degrees in computer science and business but no formal education in clinical or medical science. Integration into this trial encompassed three active roles (see Figure 5.2). I played a key

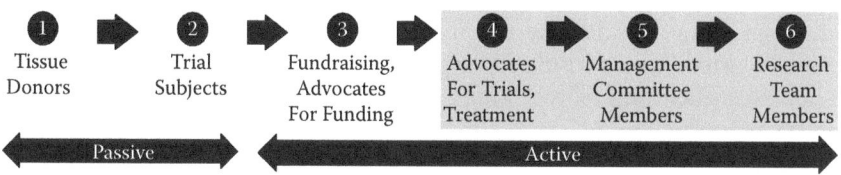

Figure 5.2 E4A03 trial active patient participation.

role in trial design as a member of the research team, in the oversight of trial as a member of Eastern Cooperative Oncology Group's (ECOG's) Myeloma Committee and Executive Committee, and, working through the International Myeloma Foundation (IMF), in educating the community about the trial to help facilitate accrual.

As the backbone of novel agent protocols, starting with thalidomide, clinicians used a high-dose steroid regimen (dexamethasone) proven in cytotoxic treatment protocols. Other novel agents followed (i.e., bortezomib, lenalidomide), improving average survival by as much as 50% by 2008 (Kumar et al. 2008). That trend continues and the pipeline for new agents remains very promising (carfilzomib and pomalidomide are two new novel agents approved for myeloma in the past year).

Research and clinical practice focused on maximum tolerated doses. Most centers went up to 200 mg/day for thalidomide, but a few centers went up to 800 mg/day. For dexamethasone (dex), there were time-honored high-dose regimens, each including 40 mg taken 4 days in a row, followed by a rest period of anywhere from 4 to 24 days.

Dex side effects were much dreaded and much discussed in the patient community. Patients taking dex were plagued by insomnia, bloating, weight gain, mania, and even instances of psychotic breaks. These were followed by withdrawal symptoms as they entered their rest periods (e.g., weakness, fatigue, low blood pressure, bone pain). Because of these severe side effects, many physicians increased their patients' rest periods, or reduced the steroid dose, or shifted to better tolerated steroids like prednisone or methylprednisolone (Solumedrol®).

Thalidomide side effects were also an issue, which included somnolence, peripheral neuropathy, deep vein thromboses (DVTs), and constipation. These side effects were dose related, resulting in many physicians reducing their patients' thalidomide dosage. Some gave as little as 50 mg every other day versus the commonly used 200 mg per day that was emerging as the standard.

In my role as patient representative serving on the ECOG's Myeloma Committee, I put forward a concept for a dose finding trial for thalidomide/dexamethasone. This concept reflected personal experience of being treated with lower than standard doses of thalidomide and dex, as well as anecdotal input from other patients in support groups and patient/family seminars organized by the International Myeloma Foundation (http://myeloma.org). My disease responded to low-dose dex (40 mg weekly), plus low-dose thalidomide (50 mg/day). Other patients reported similar successes, including some taking even lower thalidomide doses (50 mg every other day).

ECOG's senior statistician reviewed the proposed schema, declaring it too complex, stating that it would require accrual of over 2,000 patients to properly power the study (10% of the estimated 20,000 new cases of

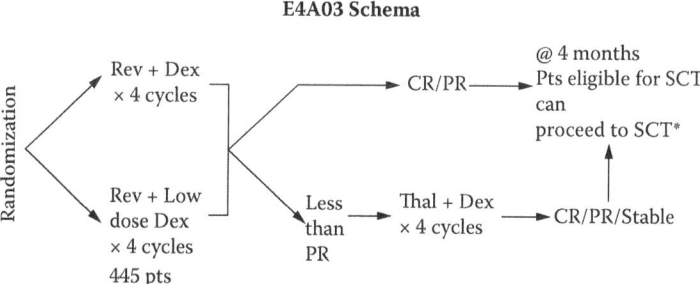

Figure 5.3 E4A03 schema.

myeloma in the United States each year) (Durie 2012). Beyond that, ECOG's Myeloma Committee was moving beyond thalidomide to focus on a promising next-generation thalidomide analog, lenalidomide (Revlimid). Following these discussions, Dr. S. Vincent Rajkumar, the chair of the Myeloma Committee, decided to include the dex dose question in his proposed design for a Phase III lenalidomide trial, E4A03 (see Figure 5.3).

The proposed trial would compare Rev (Revlimid) plus standard-dose dex to Rev plus low-dose dex. Patients enrolled in the low-dose arm would receive one-third of the dex dose of those in the standard-dose arm (160 mg versus 480 mg), with dex being given at a dose of 40 mg once weekly instead of the 4-day regimen with 4-day rest periods (see Figure 5.4).

The trial design was controversial for two reasons: First, phase III trials for new agents typically use a single agent (e.g., dex) as the standard care/comparator arm. The experimental arm would then include the standard (dex) plus the new agent. This trial design included the new agent

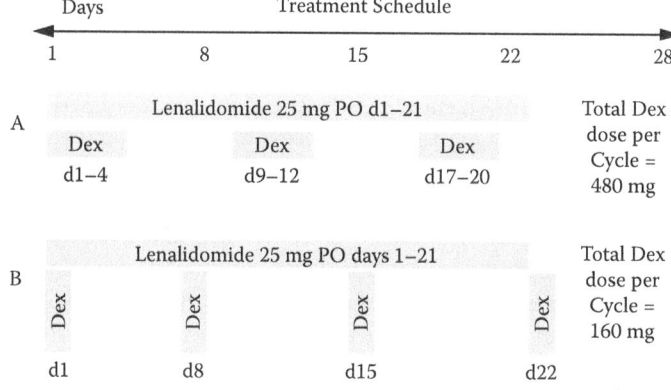

Figure 5.4 E4A03 treatment schedule.

in both arms. Hence, there could be no comparison of response for novel-agent-treated patients to those not receiving the novel agent (Revlimid). Second, oncologists and hematologists treating cancer normally use the maximum tolerated dose (MTD) of each agent to get the deepest possible response. The proposed dex dose in the experimental arm (B) was one-third the established MTD that had been the standard of care for decades.

These objections were prominent in the ECOG Myeloma Committee's discussion of the proposed schema and treatment plan. One memorable comment stands out in my memory: "Patients need to accept the fact that cancer treatments have side effects." Fortunately, with the strong support of the Dr. Rajkumar and a few outspoken committee members, the proposed trial was approved by ECOG, by Celgene, and ultimately by the National Cancer Institute's Cancer Therapy Evaluation Program (CTEP). There was significant controversy in the scientific community about the trial design. One article in a peer-reviewed journal likened the E4A03 design to "the emperor's new clothes" (Ballester 2008).

E4A03 was very attractive to the patient community. It offered the chance to use a new, novel agent (Revlimid) that was similar to thalidomide and shown in early trials to have fewer side effects than thalidomide. It was also shown to be effective in many patients whose disease had become refractory to thalidomide. Revlimid is an analog of thalidomide, engineered to avoid thalidomide's teratogenesis (birth defects), constipation, neuropathy, and somnolence. The new/novel agent was given to all E4A03 participants. And, randomization offered the chance to receive the lower dose dex regimen. Finally, the trial also offered an all-oral regimen, meaning fewer clinic visits and potentially fewer needle sticks. The International Myeloma Foundation filmed an educational video about the trial and publicized the trial on its website.

A planned interim analysis showed that one-year survival was demonstrably better in the experimental, low-dose dex arm. Overall survival was 96% (95% CI 94–99) in the low-dose dexamethasone group compared with 87% (82–92) in the standard-dose group (p = 0.0002). As a result, the trial was halted and patients on standard-dose therapy were crossed over to low-dose therapy (Rajkumar et al. 2010). The results of this trial were ultimately used to support reducing the dex dose specified on the FDA labeling.

Although E4A03 showed the superiority of lower dose dex in combination with lenalidomide for newly diagnosed myeloma patients, it changed the standard of care for other disease stages (e.g., relapsed/refractory) as well as for the use of dex in combination with other agents. Low-dose dex became the global norm in dex-containing regimens for multiple myeloma.

There are a number of factors contributing to the success of the patient-driven changes to the E4A03 trial design:

1. I had over a decade of experience with the ECOG Myeloma Committee and had strong working relationships with the chair (Dr. Rajkumar) and other committee members.
2. Clinicians had been using dose modifications for many of their patients who were having issues tolerating high-dose dex in combination with novel agents.
3. Patients were aware of others' experiences with dex dosing, owing to participation in face-to-face support groups, listservs (email-based chat groups), and educational programs.
4. Dr. Rajkumar, the committee chair, became an advocate for the patient's position in the committee's deliberations.
5. Pharma (i.e., Celgene) was supportive because lowering the dex dose would strengthen Revlimid's position versus other available (non-Celgene) treatment regimens, some of which did not include dexamethasone.

Without my involvement in the management team and the courage of the committee chair to move forward with an unorthodox trial design, myeloma patients would still be dealing with the severe side effects and higher mortality of high-dose dex.

Case study 3: Patient leads effort using consumer focus groups to optimize eligibility and randomization criteria for the TailoRx breast cancer trial

Patient advocates have been involved at ECOG for over 15 years. Some of the advocates, including breast cancer survivor Mary Lou Smith, have been involved from the outset. During her tenure, Smith gained a wealth of knowledge and experience, forging strong relationships with the senior investigators. These factors proved instrumental in this success story, where patient-led focus groups brought a wealth of data from the breast cancer community to create a patient-friendly trial design (i.e., one that patients would consider an attractive option).

Smith's involvement encompassed three active roles, owing to her work as an advocate, a member of the ECOG executive committee, and a member of the ECOG Breast Core Committee.

Node negative, estrogen-receptor positive breast cancer patients completing surgery and, in some cases, radiation treatments, are routinely given adjuvant hormonal therapy and, in most cases, chemotherapy. The recurrence rate for patients who do not receive chemotherapy is estimated to be 15% to 20%. Chemotherapy lowers the risk of recurrence by 5%, to 10% to 15% (i.e., a 25% to 33% reduction in the number of recurrences). Thus, 80% to 85%

of the patients are not helped by chemotherapy because the disease recurs in only 15% to 20% of patients. So, the benefit of chemotherapy accrues to the 5% who avoided a recurrence because of the chemotherapy. But, in order to get this benefit, 80% to 85% of the patients who would not have a recurrence are being treated with chemotherapy (Sparano and Paik 2008).

TailoRx (*Trial Assigning IndividuaLized Options for Treatment/Rx*) is the first trial from the National Cancer Institute Program for the Assessment of Clinical Cancer Tests (PACCT). The objective of the trial is to determine if using the Genomic Health 21-gene assay (Oncotype DX™) to assess the risk of recurrence will enable physicians to identify those patients who would benefit from adjuvant chemotherapy. The trial design is complex (see Figure 5.5) and its accrual target ambitious (~10,000 patients.) Patients with a low Oncotype DX recurrence risk score (RS) were assigned to arm A, which entails hormonal therapy. Patients with a high RS were assigned to arm D, to receive hormonal therapy and chemotherapy. Patients with a moderate RS (11–25) are randomized to either arm B (hormonal therapy) or arm C (hormonal therapy plus chemotherapy.)

The TailoRx trial closed to accrual after accruing 11,000 patients in October 2010 (surpassing its goal of 10,000). It is now following these patients to determine the benefit, if any, of chemotherapy for patients with moderate risk scores (i.e., RS 11–25). When the trial design was being finalized, there were concerns that it would be difficult to accrue such a large number of participants.

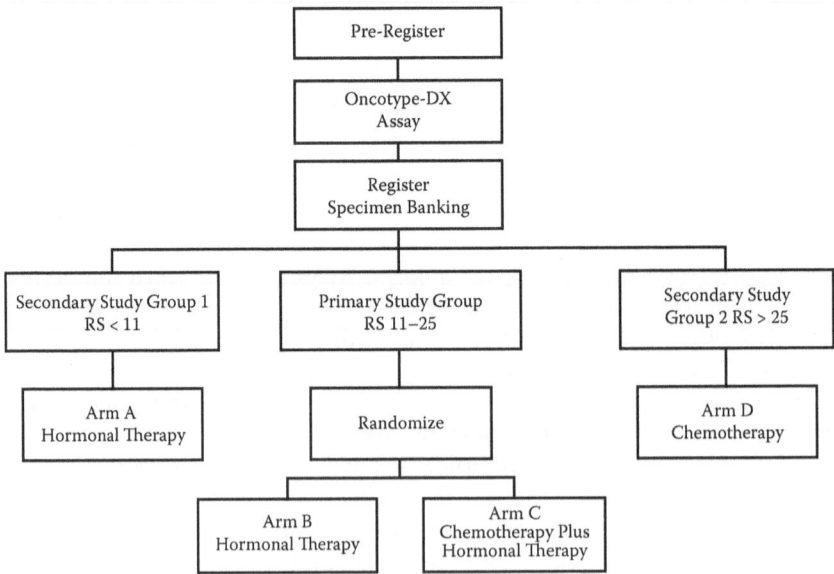

Figure 5.5 TailoRx study schema.

Smith cochairs the ECOG Patient Representative Committee and is a member of ECOG's Breast Core Committee. She is also a past president of the Y-Me Breast Cancer Organization (Y-Me) and a cofounder of the Research Advocacy Network (RAN). Working with TailoRx's principal investigator, Dr. Joseph Sparano, Smith brought to bear the expertise and resources of Y-Me and RAN to address the accrual concern. The result was a successful proposal to conduct patient and advocate focus groups to provide input to the research design and provide information to be used for patient education about the trial (Smith, Railey, and Perotti 2005).

George Sledge, MD, is past president of the American Society for Clinical Oncology (ASCO) and professor of medicine and pathology at the Indiana University School of Medicine. He is also a past chair of the ECOG breast committee. Sledge said of the effort: "This was the first time a cooperative group used focus groups during the design stage. It influenced the design and informed the conversation about eligibility and how the trial would be presented to potential participants. It (the market research) did have an effect. Not on the basic question but on how we thought about the design. We broadened our criteria and became more realistic about our accrual goals."

With Smith's perspectives on patients' issues in choosing adjuvant breast cancer treatment, with the support of Sparano and Sledge for Smith's proposal, TailoRx exceeded its accrual goal and is on track to answer the adjuvant chemotherapy question.

Case study 4: Patient participation in FDA Oncology Drug Advisory Committee (ODAC) as committee member and as speakers at new drug application (NDA) meeting influences approval decision

When pharmaceutical companies file new drug applications (NDAs) for FDA approval, the FDA can make a decision to approve the drug without an Oncology Drug Advisory Committee (ODAC) hearing. It did this recently for pomalidomide (Pomalyst™), which was approved in February 2013 for treatment of multiple myeloma based on positive clinical trial results (Vij et al. 2012).

When Onyx Pharmaceuticals filed its NDA for the proteasome-inhibitor carfilzomib (Kyprolis™), the FDA had concerns about trial participants who had cardiovascular and respiratory issues while on study. It therefore convened a meeting of ODAC to review the concerns and make a recommendation to approve or disapprove. James Omel, MD, a myeloma patient diagnosed over 15 years ago, served as ODAC's Patient

Representative for the carfilzomib meeting. Omel has extensive advocacy experience, including 4 years of service on the NCI's Board of Scientific Advisors and membership on NCI's Myeloma Steering Committee.

At the ODAC meeting, Onyx presented its trial results and perspective on approval, including presentations addressing:

- The unmet need in multiple myeloma that carfilzomib would address
- Clinical efficacy results
- Clinical safety
- Benefit–risk summary

The FDA presented its observations about the NDA, including concerns regarding safety, citing the cardiovascular and respiratory problems seen in some of the patients enrolled in the trial. Omel characterizes the tone of ODAC members' questions to the presenters as very negative about the drug, probing hard on the safety issues. The presentations were followed by an "open mike" segment where 12 people in the audience, 8 of them myeloma patients or caregivers, provided their input. The patients spoke passionately about the unmet need for new treatment options for those who have run out of options. Omel and others made the case that relapsed/refractory patients are willing to accept reasonable risk, as their options are limited and they face death if they do not have a viable treatment option.

Omel observed that the tone of the meeting changed dramatically after the patients' voices were heard. Patients drove home their views of the benefit–risk trade-off and ODAC and FDA listened. Carfilzomib was approved for treatment of patients with relapsed and refractory multiple myeloma who have received at least two prior lines of therapy. The vote was 11 to 0, with 1 abstention.

Without the patient voices in the room, without the presence of patients with long track records in advocacy and research, it is unclear if the drug would have been approved. Experienced patients and advocates bring to bear expertise about current treatments and unmet needs. And, they are able to coherently and articulately represent their constituencies and communicate their views in a public forum like an ODAC meeting.

Case study 5: Advocacy organization addresses postmarketing safety issues with a next-generation bisphosphonate

This is another personal story in which I, the author, am the protagonist.

Bisphosphonates have long been a staple of supportive care for cancer-related bone disease. Bone lesions are particularly problematic in metastatic cancers (e.g., breast cancer) and are a common symptom of active multiple myeloma. Novartis' pamidronate (Aredia™) was routinely given monthly as supportive care for multiple myeloma and patients with solid tumor bone metastases. The Aredia patent expired in May 2001. Novartis' next-generation bisphosphonate, Zoledronate (Zometa™) was approved in February 2002. After the Zometa approval, Novartis shifted its marketing focus from Aredia to Zometa. The Zometa approval specified a 15-minute infusion time versus 4 hours for Aredia.

Beginning just a year after the Zometa approval, the International Myeloma Foundation (IMF) began receiving anecdotal reports of dental problems from patients being treated with Zometa. These reports were received via their telephone hotline as well as in-person and online support groups.

Patients were reporting erosion of bone around teeth (i.e., loose teeth), exposed jawbone in their mouths, and jaw pain. Maxillofacial surgeons Marx (2003) and Ruggiero (2004) reported patients treated with bisphosphonates developing osteonecrosis of the jaw (ONJ). Ruggiero observed that all of the patients with ONJ had been treated with bisphosphonates and that number of cases at Ruggiero's facility (63 over a 3-year period) was unusual, as the prior observed frequency was just 2 per year. These findings caused great concern as bisphosphonates were widely used in myeloma and solid tumor metastases (e.g., breast) and thought to be safe.

Novartis initiated a chart review of patients at MD Anderson Cancer Center, looking back at 963 patients (including 631 breast cancer and 148 myelomas) (Novartis Pharmaceuticals Inc. 2005). The chart review looked at bisphosphonate-treated patients going back 10 years, whereas the spike in ONJ cases took place in the 2 to 3 years subsequent to the Zometa approval.

In parallel, the IMF initiated a Web-based survey to gather data aimed at understanding the cause(s) and prevalence of ONJ. The survey encompassed patients' disease and treatment histories, ONJ diagnoses, and ONJ symptoms (termed SONJ, suspicion of undiagnosed ONJ). Partnering with the breast cancer advocacy group Y-Me and listserv operator ACOR (Association of Cancer Online Resources), the IMF received 1,203 completed surveys within 30 days. Of the 1,203 patients, 75% were breast cancer patients and 25% myeloma patients (see Figure 5.6).

Analysis of the survey data significantly elevated risk of ONJ among bisphosphonate-treated patients. It showed that the risk increased with duration of treatment. The data also showed increased risk for those taking Zometa versus Aredia, likely resulting from the greater potency of Zometa (see Figure 5.7). Greater risk of ONJ was also associated with tooth extractions and other invasive dental procedures. There was controversy about the validity of the results, owing to the use of patient-reported outcomes, the anonymity of the respondents, and the lack of random

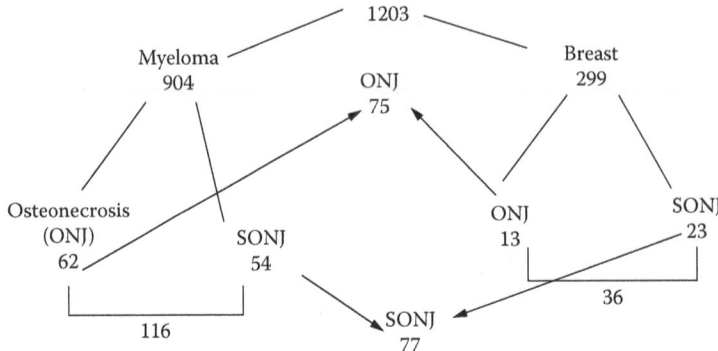

*SONJ: Suspicious findings: bone erosions; bone spurs; exposed bone

Figure 5.6 ONJ survey respondents. (With permission.)

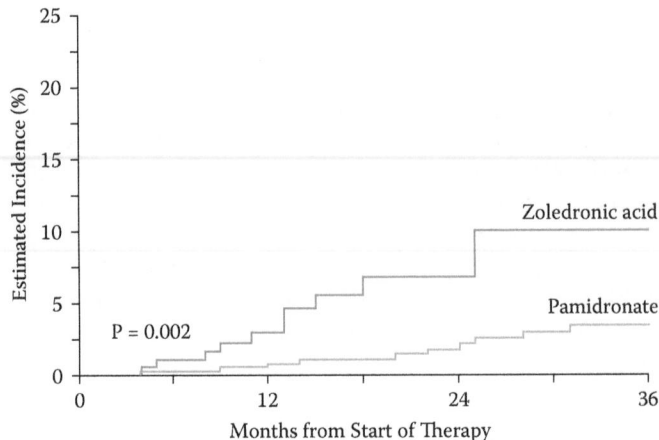

Figure 5.7 Time-dependent risk of ONJ for bisphosphonate users. (With permission.)

sampling. Despite these issues, peer reviewers accepted the study for oral presentation at the American Society of Hematology (ASH) for publication in the *New England Journal of Medicine* (Durie et al. 2005).

Initiatives by the IMF and American Dental Association to inform the clinical and patient communities about ONJ were instrumental in changing clinical practice:

- Paying closer attention to dental health when initiating bisphosphonate treatment
- Avoiding invasive dental procedures

- Limiting bisphosphonate usage to patients with active bone disease or a history of bone disease
- Limiting the duration of bisphosphonate treatment to 2 years and moving from monthly infusions to 3- to 6-month intervals

Because of the statistical issues with the Web-based survey, ONJ causality was never proven. Yet, it spurred changes in clinical practice and patient behavior that lowered the observed incidence rate from an estimated 6% of bisphosphonate-treated patients to less than 1% (Dimopoulos et al. 2009). Ironically, because of a now low incidence rate, causality will likely never be proven. Nevertheless, the outcome is positive because fewer patients will be stricken with ONJ.

Without the rapid feedback from the patient communities gathered by the IMF and Y-Me, the issue of ONJ as a risk factor of bisphosphonate use could have taken years to come to the forefront. Without the Web-based survey and the subsequent publications, speaking engagements, and educational programs, many more people would have been diagnosed with ONJ. Action by advocacy organizations, using their strong ties to patient communities, clinicians, and scientists, helped resolve a serious postmarketing safety crisis.

Moving forward to more broadly involve patients

Advocates and advocacy organizations have established track records working as advocates for research funding and fundraising for research, awareness, and participation in clinical trials, as well as education about available treatment options. Advocates and advocacy organizations also serve as information clearinghouses for the scientific, clinical, and patient communities. The IMF, via its International Myeloma Working Group (IMWG), has driven publication of consensus guidelines to address the many situations where there is no clear evidence-based standard of care ("International Myeloma Working Group Projects" n.d.). Via online communities, support groups, and educational programs, advocacy organizations provide communication channels for identifying emerging treatment trends (e.g., off-label usage, dose reductions) and safety issues.

There remain mixed feelings within the scientific and pharma communities about patient involvement. There are those that treat patient involvement as a "necessary evil," like audits and institutional review boards (IRBs). To be sure, the impact of patient involvement depends heavily on the experience, knowledge, and capabilities of the patients involved. Potential impact also depends on the nature of the research and the dynamics of the research team. The case studies presented in this chapter show that with the right people and the right situation, patient involvement can yield spectacular results, as shown in Table 5.3.

Table 5.3 Case Study Impact Summary

Case study	Impact
1. Bringing thalidomide to myeloma	Successful introduction of the first novel agent for treatment of multiple myeloma
2. Establishing low-dose dex standard	Changing the global standard of care for use of dexamethasone in multiple myeloma
3. Focus groups inform trial design	Successful accrual of over 11,000 patients to answer a critical question about breast cancer treatment
4. Patient voices argue for approval at FDA panel	Approval of a new therapy for multiple myeloma
5. Advocates address postmarketing safety issue	Dramatic reduction in incidence of bisphosphonate-related osteonecrosis of the jaw

Like the research process, patient involvement is seldom a source of immediate gratification. I recall my first involvement with reviewing and scoring research grant applications. It was hard to be enthusiastic about the many basic science proposals because they seemed so far from anything that might ever translate into a treatment. Clinicians and scientists participating in the peer review patiently explained that what we learn through basic scientific research eventually translates into clinical research that leads to better patient outcomes.

So it is with patient involvement. When patients first become involved, they face substantial challenges, dealing with scientific and medical jargon, learning about the research process and the state of the science. Patients integrated into research teams typically face challenging personal dynamics and a formidable learning curve. Like research results, the results of patient involvement vary, with both failures and successes to be expected. And, it takes time to realize the potential. But, you've got to be in it to win it.

References

Ballester, O. 2008. "The Emperor's New Clothes or the Current Practice of Clinical Trials for Multiple Myeloma in the USA." *Cancer Investigation* 26(5):445–447.

Celgene Corporation. 2011. "Annual Report, Form 10-K," p. 52.

Dimopoulos, M. A. Kastritis, E., Bamia, C., Melakopoulous, I., Gika, D., Roussou, M., Miglou, M., et al. 2009. "Reduction of Osteonecrosis of the Jaw (ONJ) after Implementation of Preventive Measures in Patients with Multiple Myeloma Treated with Zoledronic Acid." *Annals of Oncology* 20(1):117–120.

Durie, B. G. M. 2012. *IMF Multiple Myeloma Patient Handbook 2012/2013*, p. 5. http://handbook.myeloma.org.

Durie, B. G. M., Katz, M. S., and Crowley, J. 2005. "Osteonecrosis of the Jaw and Bisphosphonates." *New England Journal of Medicine* 353:99–102.

"International Myeloma Working Group Projects." n.d. *Wikipedia*, http:// en.wikipedia.org/wiki/International_Myeloma_Working_Group#Projects.

Kumar, S. K., Rajkumar, S. V., Dispenzieri, A., Lacy, M. Q., Hayman, S. R., Buadi, F. K., Zeldenrust, S. R., et al. 2008. "Improved Survival in Multiple Myeloma and the Impact of Novel Therapies." *Blood* 111(5):2516–2520.

Marx, R. E. 2003. "Pamidronate (Aredia) and Zoledronate (Zometa) Induced Avascular Necrosis of the Jaws: A Growing Epidemic." *Journal of Oral and Maxillofacial Surgery* 61(9):1115–1117.

"Myeloma Today Profile: Seema Singhal, MD." 2001. *Myeloma Today* 4(7). http:// myeloma.org/ArticlePage.action?articleId=483.

National Institutes of Health. 2008. "COPR Role of the Public in Research Working Group." http://www.nih.gov/about/copr/reports/documents/10312008_RPR.pdf.

Novartis Pharmaceuticals Inc. 2005. "Background Information for Oncologic Drugs Advisory Committee Meeting." March 4.

Rajkumar, S. V., Jacobus, S., Callander, N. S., Fonseca, R., Vesole, D. H., Williams, M. E., Abonour, R., et al. 2010. "Lenalidomide Plus High-Dose Dexamethasone versus Lenalidomide Plus Low-Dose Dexamethasone as Initial Therapy for Newly Diagnosed Multiple Myeloma: An Open-Label Randomized Controlled Trial." *Lancet Oncology* 11(1):29–37.

Ruggiero, S. L. 2004. "Osteonecrosis of the Jaws Associated with the Use of Bisphosphonates: A Review of 63 Cases." *Journal of Oral and Maxillofacial Surgery* 62(5):527–534.

Singhal, S., Mehta, J., Desikan, R., Ayers, D., Roberson, P., Eddlemon, P., Munshi, N., et al. 1999. "Antitumor Activity of Thalidomide in Refractory Multiple Myeloma." *New England Journal of Medicine* 341:1565–1571.

Smith, M. L., Railey, E., and Perotti, J. 2005. "Use of Focus Groups to Inform Clinical Trial Design." The Department of Defense (DOD) Breast Cancer Research Program (BCRP) 6th Era of Hope Conference, Orlando, Florida.

Sparano, J. A., and Paik, S. J. 2008. "Development of the 21-Gene Assay and Its Application in Clinical Practice and Clinical Trials." *Clinical Oncology* 26(5):721–728.

Vij, R., Richardson, P. G. G., Jagannath, S., et al. 2012. "Pomalidomide (POM) with or without Low-Dose Dexamethasone (LoDEX) in Patients (pts) with Relapsed/Refractory Multiple Myeloma (RRMM): Outcomes in pts Refractory to Lenalidomide (LEN) and/or Bortezomib (BORT)." *Journal of Clinical Oncology* 30(Suppl; abstr 8016).

Zimmer, C. 2010. "Answers Begin to Emerge on How Thalidomide Caused Defects." *New York Times*, March 15. http://www.nytimes.com/2010/03/16/science/16limb.html.

chapter six

Innovative organizations
Viewpoints of organizational scholars and practitioners

Lindsey Kotrba, Ia Ko, and Daniel R. Denison

Contents

Innovation matters to every organization regardless of the industry. Businesses today need to succeed within a highly competitive, rapidly changing, and increasingly global marketplace (e.g., Hurley and Hult, 1998; Olausson and Berggren, 2010; Tuominen, Rajala, and Möller, 2004). In response to these challenging operating environments, innovation has been identified as a critical source of competitive advantage for organizations (e.g., Bouncken and Kraus, 2013; Miron, Erez, and Naveh, 2004; Tellis, Prabhu, and Chandy, 2009; Van de Ven, 1998). Empirical research supports this assertion, linking organizations' innovation effectiveness to overall organizational performance (e.g., Damanpour, Szabat, and William, 1989; Kotler, 1991; Subramanian and Nilakanta, 1996). The relationship between organizational innovation and indicators of organizational performance is positive and robust, and has been demonstrated in many different organizational contexts (Droge, Calantone, and Harmancioglu, 2008; Han, Kim, and Srivastava, 1998). There is no doubt, as past research has continually suggested, that organizations benefit from being innovative.

For organizations in knowledge-intensive industries, such as pharmaceuticals and consumer electronics, innovation becomes the driver of their survival and success (Sorescu, Chandy, and Prabhu, 2003). However, pharmaceutical firms are faced with a laundry list of innovation challenges. Decreasing productivity, patent expiry, rising costs of research and development (R&D), high attrition rate of compounds in phase 2, high regulatory hurdles, increasing concern about adverse side effects, and so on; the list covers a wide range of ongoing challenges. Current innovation challenges in the pharmaceutical industry are well documented elsewhere (e.g., Comanor and Scherer, 2013) and discussing them in detail is beyond the scope of this chapter. Instead, we try to bring an organizational scholar and practitioner perspective on the topic of innovation. And as such, this chapter will cover some of the innovation lessons we have learned from our research and experience working with various organizations across multiple industries. By doing so, we hope to provide pharmaceutical firms with insight around good innovation habits to be preserved, bad innovation habits to change, and help the industry rethink how it innovates by learning from innovators in other industries.

Understanding and measuring innovation

Innovation is an extraordinarily broad topic and the term itself has been defined in numerous ways. However, in line with Van de Ven and Angle (1989), we define innovation as the process of bringing a new idea into use. Fleshed out, the innovation process involves a multistep process of generation, acceptance, and implementation of new ideas. For example, within new product development (NPD) functions of businesses, innovation may reflect a stage-gate process including iterative progression through several phases, from scoping to building a business case, development, testing and evaluation, and, finally, launch (Cooper, 2011). Beyond implementation (or launch), as Medina, Lavado, and Cabrera (2005) point out, several authors have suggested that a new idea must also be successful once implemented to be considered an innovation (e.g., Burgelman and Sayles, 1988; Cumming, 1998; Guellec, 1999). Similarly, progression through a stage-gate innovation process assumes sufficient stakeholder support (i.e., success) at prior stages. Therefore, we too adopt this notion and recognize that the concept of success is built into the innovation process. That is, innovation includes both creative idea generation and successful implementation of those creative ideas. Although creativity is important, the test of innovation is whether the innovation contributes to market and customers, not whether it is scientifically or technologically important (Drucker, 1999). This is also the case in the pharmaceutical industry. The success of innovation includes both technological advancement and market attractiveness (Plotnikova, 2010).

In general, product innovations can be thought of as existing in four groups based on two dimensions: the extent of advancing technology and the extent of outperforming existing products in fulfilling customer needs (Chandy and Tellis, 1998). A *radical innovation* is high on both dimensions; a *technological breakthrough* is high on the first dimension only; a *market breakthrough* is high on the second dimension only; and an *incremental innovation* is low on both dimensions. This framework can also be applied to describe and classify pharmaceutical innovations. New molecular entities (NMEs) are high on the technological advancement dimension as they are based on a new active ingredient. NMEs granted a priority review by the U.S. Food and Drug Administration (FDA) would be considered radical innovation as that means the drug is perceived to advance available therapy (i.e., customer benefit) as well as the current technology (Sorescu et al., 2003). NMEs given a standard review would be classified as a technological breakthrough. Non-MNEs given a priority would be considered a market breakthrough since they have high therapeutical potential although they are based on a new usage, formulation, or dosage of existing components. Radical innovations and breakthroughs are rare. They represent only about 7% of all new drugs (255 breakthroughs out of 3891 new drug applications from 1991 to 2000; Sorescu et al. 2003).

Aside from defining innovation, the issue of how to measure the construct presents many challenges (McLean, 2005; Tuominen et al., 2004). Innovation success has been measured in various ways including, but not limited to, subjective assessment, patents, market entry, and organizational growth (e.g., Gopalakrishnan and Bierly, 2006; Jalles, 2010; Rosenbusch, Brinckmann, and Bausch, 2011). In the pharmaceutical industry, the FDA (2012) releases its annual list of innovative drugs covering an array of new products entering the marketplace. Although market entry is one good indicator of innovation success, not all new drugs (or products in general) entering the market reflect the same innovation success. Aforementioned, innovation includes both creative idea generation and successful implementation. Some innovations are novel and yet reap marginal financial benefits, whereas others obtain bigger market gain despite their lack of novelty.

One unique aspect and perhaps one of the biggest challenges of pharmaceutical innovation is that the overall process has a long-term time horizon and requires various resources. Figure 6.1 illustrates the innovation process in the pharmaceutical industry as well as a more general innovation process. The innovation process in the pharmaceutical industry usually entails at least three large stages: drug discovery, development, and FDA review (and then market entry). Each of these stages is further broken down to multiple processes. For instance, the drug discovery stage

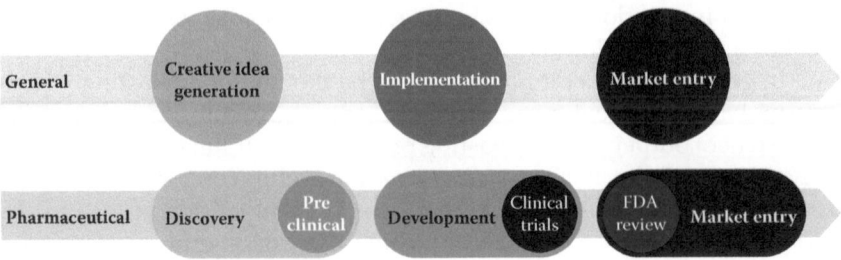

Figure 6.1 Innovation process.

starts with prediscovery, proceeds to target identification and validation, and then finally to preclinical testing. The overall process of drug discovery, development, and FDA review takes a long time, spanning 10 to 15 years, and involves a large number of people from multiple disciplines. Many of the drugs that become available to patients recently were in the discovery stage before the year 2000.

What innovative organizations have in common

When it comes to driving innovation, the existing literature offers insight into the characteristics that innovative organizations most commonly share (e.g., Medina, Lavado, and Cabrera, 2005). A great deal of research has focused on how organizational structure, job characteristics, and management processes relate to creativity and innovation (e.g., Amabile, 1996; Axtell, Holman, Unsworth, Wall, and Waterson, 2000; Bonner, Ruekert, and Walker Jr., 2002; Calantone, Harmancioglu, and Droge, 2010; Damanpour, 1991; Oldham and Cummings, 1996; West, Smith, Feng, and Lawthom, 1998).

Structure of innovative organizations

First, several studies have investigated structural variables as determinants of innovation (e.g., Damanpour, 1991; Wolfe, 1994). The majority of scholars and practitioners claim organizations with flexible and organic structure are more likely to innovate (e.g., Bonner et al., 2002; Pitt and Clark, 1999). However, research findings have been mixed and indicate the role of structure in innovation is about balancing formalization and flexibility rather than simply adopting an organic structure (Calantone et al., 2010). Creative organizations tend to have flexible structures with few rules and regulations, and loose job descriptions (Andriopoulos, 2001). In organic and flexible structures, there is more autonomy and resource flexibility and this can facilitate more fluid creative idea generation and innovative solutions to a problem. Yet in mechanistic structures, a high level of formality provides a sense of structure and reduces ambiguity, and this formality of execution can promote more effective cross-functional collaboration and coordination and lead to more successful implementation of creative ideas (Olson, Slater, and Hult, 2005; Tatikonda, 1999; Tatikonda and Rosenthal, 2000). Thus, both can play a role in innovation.

The balancing between formality and flexibility might depend on the type of innovation. Menguc and Auh (2010) found that mechanistic organizational structures were associated with successful implementation of *incremental* product innovations, whereas organic organizational structures were linked to better new product performance of *radical* innovations. So in sum, it is unlikely that there is one best organizational structure for driving successful innovation. But as a general guide providing as much flexibility as possible within a thoughtful structure can promote both creativity and ensure successful execution. Successful organizations are those that have learned how to strike the right balance between flexibility and stability.

Organizational size and innovation

Uncovering the relationship between firm size and innovation effectiveness has long been a topic of investigation for many innovation scholars. Although the answer may not be as simple as "the bigger, the more innovative" or "too big to innovate," size is linked to innovation. Dominant firms (i.e., firms with bigger market share, assets, and profits) are more likely to be equipped with technological, financial, and market-related resources and thus handle risks associated with innovation better (Sorescu et al., 2003). With economies of scale and scope in R&D and marketing, large firms tend to produce not only more radical innovations but also technological and market breakthroughs (Chandy and Tellis, 2000). In the pharmaceutical industry, dominant and large firms not only introduce most innovations, especially radical innovations and breakthroughs but also tend to gain more (i.e., higher new product value) largely due to their capability to provide product support in technology and marketing (Sorescu et al., 2003).

Small firms might be slightly disadvantaged when it comes to innovation as they often lack resources for funding high-risk research (Gnyawali and Park, 2009) or for buffering them from failures (Bougrain and Haudeville, 2002). However, their advantage is that they are more nimble and can make quick changes in response to market demands (Andriopoulos and Lewis, 2009). Research shows that small firms with successful innovation records often focus on basic research, drug discovery, and preclinical experiments rather than on development. Also, small and medium firms may optimize their capability to innovate through high-quality patents by focusing on faster learning and developing a narrow knowledge base (Gopalakrishnan and Bierly, 2006). Overall, research on the relationship between firm size and innovation highlight the importance of understanding the implications of firm size on innovation. Larger companies might dominate, but they face some costs associated with being big; smaller companies have resource challenges but can enhance their innovation success by using innovation strategies that are appropriate for their size.

Intrafirm collaborations

In addition, innovative organizations often benefit from intraorganizational collaborations (Faems, Van Looy, and Debackere, 2005). Through partnerships with other organizations (and sometimes universities), companies gain various benefits that facilitate innovation such as having better access to information, knowledge, skills, capabilities, experience, and technology; sharing resources; and reducing costs (Hotz-Hart, 2000). Traditionally, pharmaceutical companies have heavily relied on their own internal R&D for innovation. In response to innovation challenges, we have more recently begun to see more intrafirm collaborations among pharmaceutical companies such as R&D collaborations (e.g., Transcelerate), open innovation (e.g., Centres of Excellence for External Drug Discovery), innovation partnerships (e.g., GSK–McLaren, Teva–Procter & Gamble), and outsourcing (e.g., Bristol-Myers Squibb).

Research shows that the success of product innovation through intraorganizational collaboration depends on the continuity of collaboration and diversity of partners (Nieto and Santamaria, 2007). It might sound obvious, but it is critical to take time to gain experience in partnership management, to learn to collaborate, and to build mutual trust (Bouchken and Fredrick, 2012). This also implies that organizations may not reap benefits from their collaborations as soon as they might hope. Furthermore, collaboration partner choices can greatly impact innovation success. In Neito and Santamaria's (2007) study, suppliers as collaboration partners showed up as having the most significant effect on product innovations, but research organizations and customers also appeared to be good innovation partners. Collaboration with competitors did not significantly impact innovation; and in fact, it had a negative effect on novel innovations.

This brings up an interesting point about collaboration with competitors, or *coopetition*. Coopetition has been recognized as an important innovation success strategy in several studies (e.g., Bouncken and Kraus, 2013; LeRoy and Yami, 2009), and the existing literature highlights both pros and cons of coopetition. Similar to other types of intrafirm collaboration, coopetition enables organizations to take advantage of additional knowledge and resources (Carayannis and Alexander, 1999;

Dubois and Fredriksson, 2008). However, as above earlier and revealed by Nieto and Santamaria (2007), coopetition might not be a good strategy for firms desiring to create highly novel innovations. Bouncken and Kraus (2013) investigated the effect of coopetition on revolutionary and radical innovation among small- and medium-sized enterprises (SMEs) in knowledge-intensive industries. Although both are forms of novel innovation, revolutionary innovations are extremely novel and involve greater technological and market discontinuities than radical innovations. They found that in general, coopetition has a negative impact on revolutionary innovation and a positive impact on radical innovation. Interestingly, coopetition was positively associated with revolutionary innovation when there was a high level of technological uncertainty—as coopetition can help organizations reduce uncertainty—and a greater degree of inlearning (i.e., internal learning of external knowledge). To sum up, for firms to maximize the benefits of intraorganizational collaborations, it is vital to ensure a diverse network of carefully chosen partners, and perhaps more important to build relationships with those partners so that they lead to the access of diverse resources to promote more novel innovations.

Organizational culture of innovative organizations

Last, an increasing number of studies provide empirical evidence supporting the link between organizational culture and innovation (e.g., Brentani and Kleinschmidt, 2004; Büschgens, Bausch, and Balkin, 2013; Lyons, Chatman, and Joyce, 2007; Martins and Martins, 2002). Organizational culture refers to values, beliefs, and assumptions held by the members of an organization and which facilitate shared meaning and guide behavior at varying levels of awareness (Denison, 1996). Research shows that innovative organizations often have a culture of entrepreneurship, openness, learning, risk taking, informality, and adaptability (Knox, 2002). Also, they reward innovation, have a shared strategic mission and vision, encourage trust relationships between managers and employees, and are customer oriented (Brentani and Kleinschmidt, 2004; Martins and Martins, 2002).

In a study of 759 public firms from 17 countries, Tellis, Prabhu, and Chandy (2007) revealed corporate culture—not government policy, labor, or capital—was the strongest driver of innovation across nations. In particular, organizations' cultural characteristics of a willingness to cannibalize resources (i.e., reduce the value of its own prior investments), future orientation, and tolerance for risk as well as practices of using product champions, innovation incentives, and internal markets (i.e., the level of internal autonomy and competition) were positively associated with radical innovation. Furthermore, a recent meta-analysis by Büschgens and colleagues (2013) shows that flexible and externally oriented cultures were associated with an organizational focus on innovation, while a hierarchical culture emphasizing control and internal orientation was negatively linked to innovation.

In terms of the link between organizational culture and innovation outcomes among pharmaceutical companies, empirical research is scarce but a few studies provide interesting findings. Dorabjee, Lumley, and Cartwright (1998) conducted a survey study in five pharmaceutical companies in the United Kingdom with the intent to identify the overall cultural characteristics of the pharmaceutical industry. Compared with innovative companies in general, pharmaceutical companies showed differences in three cultural elements important to innovation: a higher level of debate, a low propensity to risk taking, and less time to work on new ideas. Also, a disagreement between leaders and nonleaders emerged such that leaders perceived their organizational culture more positive and creative than employees. Vitols (2002) discusses how management culture influenced the way "the Big Three" German chemical/pharmaceutical companies (i.e., BASF, Bayer, Hoechst) responded to pressures for change from capital markets. Although Vitols does not directly address how organizational culture impacted innovation success, the author provides insight into how culture might impact innovation strategy and future innovation success. In a more recent study, Tollin (2008) interviewed 26 marketing executives in pharmaceutical and FMCG (fast moving consumer goods) companies. Radical product innovation emerged as a key issue among the interviewees, and company culture—especially developing a more market-orientated company culture—was stated as an important element of innovation. In summary, although largely conducted outside of the pharmaceutical industry, research on culture and innovation shows a strong, clear link between the two. An increasing number of organizations are now considering organizational culture as a key driver of innovation success and exploring ways to develop and manage an innovative culture. In a later section of this chapter, we revisit organizational culture and provide a framework around different ways organizations may foster innovation through culture.

Thus far, we provided a summary of how innovative organizations are characterized in terms of their structure, size, ability to collaborate with other organizations, and organizational culture. Although research findings do not necessarily provide a very clear answer in regard to finding one best structure, size, collaboration strategy, or culture, there are common characteristics that are shared by innovative organizations. They provide enough flexibility to encourage creativity but use their formal structure and processes to facilitate innovation implementation. Big size organizations achieve innovation success by utilizing their resources and by providing as much product support as they can, whereas smaller organizations maximize their innovation capacity by developing a deeper knowledge and having a focused innovation area. Innovative organizations know with whom and how to collaborate so they can gain greater resources and knowledge. Finally, organizations with an adaptable, externally oriented, and collaborative culture are more likely to innovate successfully. Next, we provide a framework for pharmaceutical companies to consider to hopefully help boost their innovation success.

Rethink what and how you innovate

There is little question that radical, revolutionary innovations fuel growth and financial success in organizations (Tellis et al., 2007). This is why so many companies invest their resources into developing the next iPod, Square, Lipitor, or whatever the "next big thing" might be. However, there are at least two risks that are involved with this approach that need to be considered. First, this mindset can limit the view of what counts as innovation. Organizations are realizing that doing something well is just as important, and sometimes even more important, as doing something new, and are revising how they innovate accordingly (Kanter, 2010). For many organizations, innovation opportunities exist not just in new product and services. Improving production processes and efficiency, revamping marketing strategies, creating new distribution channels, or offering a better customer experience around existing products and services, are all ways that organizations can successfully innovate. Although new drug development will remain a fundamental innovation activity for pharmaceutical companies, these organizations can still create impactful innovation in other areas by looking for unmet needs of the various stakeholders (internal and external) and exploring new ways to brand, market, and sell existing drugs.

Another risk associated with focusing foremost on finding the next big idea or blockbuster innovation is that this strategy can lead to compromising the development of small ideas that can result in market breakthroughs and high revenue potential. Big ideas require a great deal of resources and often a longer time horizon. And in the end, big blockbusters are rare, and this approach can be risky, especially for small firms that lack resources that can buffer them from failures.

Innovation pyramid

Kanter (2006) suggests using an innovation pyramid approach as one remedy for this common innovation strategy conflict (Figure 6.2). The innovation pyramid has three tiers. On the top, big bets are placed on a few innovation ideas that provide clear direction for the future. These ideas receive the highest organizational priority and a good share of the resources. Promising ideas that have somewhat clear future directions lay in the middle of the pyramid. They are supported and developed by

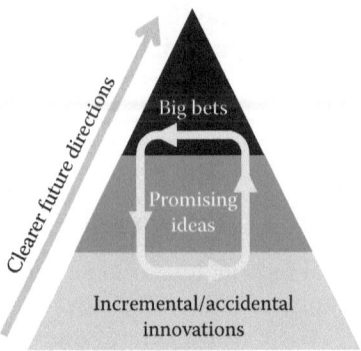

Figure 6.2 Innovation pyramid.

dedicated teams and are not yet considered organizational priorities. The bottom of the pyramid represents a broad base of early stage ideas with the potential for radical innovations or ideas for incremental innovations.

The innovation pyramid approach provides organizations with a framework to assess current innovation efforts: Are there top priority innovation activities at the organizational level (top of the pyramid)? Are there project teams developing ideas and turning those ideas into reality (middle)? Is everyone else in the company invited to contribute their ideas (bottom)? Furthermore, this approach encourages organizations to grow innovation culture by supporting innovation activities at each level and allowing room for early-stage ideas to surface and develop. Ideas at the bottom of the pyramid might seem small or unrealistic. However, small and incremental innovations can add up to big profits. Radical, revolutionary innovations often start with ideas that often seem unrealistic and fuzzy. The importance of these innovation activities occurring before the formal NPD process, or front-end innovation (FEI), has been highlighted (e.g., Koen et al., 2001). Some state improving the FEI process is the most cost-effective way to improve innovation (e.g., Backman, Börjesson, and Setterberg, 2007; Williams, Kochhar, and Tennant, 2007). To identify how radical FEI might be supported in pharmaceutical companies, Aagaard and Gertsen (2011) conducted an in-depth case study of H. Lundbeck A/S and studies of seven European and American pharmaceutical companies. They revealed 11 key factors for facilitating and supporting radical FEI. The three most important factors were team culture tolerant of failure (i.e., a blame-free culture where teams feel safe to explore ideas and try new things), efficient cross-functional and cross-disciplinary knowledge sharing and collaboration, and empowerment of employees to learn and explore.

Both Kanter's (2006) approach and research findings on FEI, especially Aagaard and Gertsen's (2011), point to the importance of building an organizational culture supportive of innovation. As mentioned earlier, organizational culture is receiving an increasing amount of attention as a key driver of innovation and is key consideration when discussing an organization's ability to innovate. As such, in the next section, through our research and experience with organizations, we explore how organizations might build a culture of innovation.

Drive culture to innovate

Innovative organizational cultures can take various forms, and building an innovative culture is not a "one type fits all" approach. Based on our research, we discuss differing organizational culture characteristics that support innovation. First, we discuss developing a culture that is focused on responding to the market as one way to drive innovation. *Market-driven* organizational cultures often have a culture that is flexible and externally oriented, and are better at listening to the marketplace and translating those demands into action. Using customers as sources of ideas for innovation, they not only "hear" but also act on the type of innovations valued by the customer. They obtain a deep understanding of customer needs by maintaining regular, quality contact with end users. Although customer-driven innovations exist in the pharmaceutical industry (e.g., market breakthroughs), our research points to customer focus as a common weakness of pharmaceutical firms. Market breakthroughs are just as good as technological breakthroughs when it comes to financial valuation (Sorescu et al., 2003), and thus pharmaceutical organizations may benefit from focusing on developing a culture that is more strongly focused on the customer. In addition, innovative companies often find solutions to challenging problems outside of their industry or field.

Other organizations have *vision-driven* cultures, and that can also be a very effective way to drive innovation. An organization's mission refers to the organization's purpose and direction, and raison d'être. If an organization's mission is not clearly established or understood, it is critical to engage everyone in the organization to define the mission and clarify how the organization creates value. A clear mission allows for the development of new ideas that are aligned with the direction in which the organization is moving, and that can be a powerful way to drive innovation. When lacking clear mission and vision, organizations tend to become distracted and end up keeping their feet in too many different efforts. With a strong mission, organizations can stay focused and say "no" to the development of new products not aligned with the organization's mission and long-term goals. To drive innovation through vision, the highest priority is to adopt a long-term strategy and direction and identify issues that have a longer time horizon. The next step is to help employees align their goals to the organization's so that everyone in the organization can define his or her own goals in terms of the organizational mission, vision, and strategy.

Last, organizations can also have cultures that drive innovation through focusing internally, on their people. For many organizations, employees are their biggest source of innovation. Google, for example, is well known for granting their employees time to work on something that interests them personally and to engage in innovation activities. Consumer goods manufacturers heavily rely on their employees for creating innovative products and reaping market successes. R&D-based innovation is one manifestation of *employee-driven* innovation, and this innovation model has been successful in the pharmaceutical industry. However, with increasing R&D costs and decreasing R&D productivity, there is a growing need for rethinking how to use R&D for innovation. Also, when we look at the most innovative companies' lists reported by various sources (e.g., strategy + business, FastCompany, Bloomberg Businessweek), the companies on these lists are not the same as the top R&D spenders.

However, an employee-driven innovation is not just about relying on designated innovators, such as R&D in pharmaceutical. To drive innovation through a focus on employee involvement, organizations need to give all individuals autonomy, resources, and opportunities to generate creative ideas and play with those ideas. Empowering employees and providing them with needed resources is vital to boosting employee-driven innovation. Also, being open to a bottom-up decision-making process allows employees to actively share their ideas and to seek to contribute to decisions. Furthermore, when organizations are built around teams, teamwork promotes ideas and information flow and encourages creativity.

Although a strong vision may be apparent when looking at Apple's culture, and Google's culture may more clearly emphasize a bottom-up approach, our research shows that the most successful organizations develop organizational cultures that do all of these things. That is, they focus on mission and strategy, listen and learn from the market and their customer, and develop and focus on their people. Different organizations may emphasize parts of these cultural characteristics over others, but if you take a close look at the cultures of the most successful innovators you will see elements of them all. And regardless of the primary way an organization fosters innovation, a consistent set of systems, processes, and practices are needed to support organizational innovation as it facilitates the implementation of creative ideas.

To foster consistency to support other strengths and fuel innovation, organizations need to facilitate coordination and integration by developing a system connecting various projects and initiatives. Creative ideas do not always result in innovation. Creative individuals need support from different parts of the organization and a system that can help make ideas reality. Innovations need connectors; surround innovators with supportive collaborators. Also, it is important to encourage knowledge flow to boost creativity so that people from different parts and levels of the organization can work well together. Furthermore, innovation success often depends on having a "strong" culture of innovation where there is more tolerance of deviance, divergent thinking, and constructive confrontation. A strong culture does not necessarily mean a high control, uniformity culture. The most innovative organizations (e.g., Southwest Airlines, Google) possess a strong culture and yet use the culture to ensure quality and consistency in producing more innovative products and services. They are highly consistent in what they do and yet stay highly innovative as well.

Implications for pharmaceutical companies

Our discussion thus far has focused on the characteristics and strategies commonly shared by innovative organizations in general. Although pharmaceutical firms face a unique set of challenges and their innovation process looks different from that of others, there are underlying innovation lessons that can be gleaned from our review. We encourage readers in the pharmaceutical industry to think about what this might mean for them and how they might drive innovation in the future.

We began this chapter with clarification on the definition and measure of innovation. Innovation can convey various meanings and be measured in different ways. Given what we know about different types of innovation and the impact organizational structure and size might have on innovation, pharmaceutical companies might clarify what innovation means for them accordingly. For instance, a small pharmaceutical company might view innovation in a more focused way such as technological breakthrough, whereas a large pharmaceutical firm might define innovation more broadly (e.g., NPD in general, new marketing process, better customer experience) and measure innovation in more than one way. By rethinking what innovation means, pharmaceuticals can prioritize *what* to innovate and *where* to innovate.

Furthermore, we explained *how* organizations might innovate. And asking how an organization innovates reveals another interesting and important point: *who* innovates? Some pharmaceutical firms may change how they innovate by collaborating with outsiders. They can partner with customers, suppliers, and even competitors to gain access to resources and maximize their innovation effectiveness. As mentioned, it is vital to carefully select who the innovation partner would be and this requires that the firm have a deep understanding of its potential collaborators. Other pharmaceuticals might alter the way they innovate by using an innovation pyramid approach and allowing more employees to contribute to innovations. They encourage everyone in the organization to innovate and provide support for innovation at various levels.

Finally, we highlighted the importance of having an organizational culture supportive of innovation—regardless of industry, firm size, organizational structure, and innovation strategy. We urge pharmaceutical firms to envision what building an innovation culture would mean for them and how they might drive vision-, market-, and employee-driven cultures to innovate.

Conclusion

In this chapter, we shared our perspectives on innovation as organizational scholars and practitioners. We covered what we have learned about innovation, innovative organizations, and innovation strategies based on the literature, our own research, and our consulting work with various organizations. We hope our chapter has provided readers with an opportunity to rethink innovation and gain insight into how they might enhance innovation success in their own organizations.

References

Aagaard, A., and Gertsen, F. (2011). Supporting radical front end innovation: Perceived key factors of pharmaceutical innovation. *Creativity and Innovation Management, 20*(4), 330–346. doi:10.1111/j.1467-8691.2011.00609.x.

Amabile, T. M. (1996). *Creativity and Innovation in Organizations.* Harvard Business School.

Andriopoulos, C. (2001). Determinants of organisational creativity: A literature review. *Management Decision, 39*(10), 834–841. doi:10.1108/00251740110402328.

Andriopoulos, C., and Lewis, M. W. (2009). Exploitation-exploration tensions and organizational ambidexterity: Managing paradoxes of innovation. *Organization Science, 20*(4), 696–717.

Axtell, C. M., Holman, D. J., Unsworth, K. L., Wall, T. D., and Waterson, P. E. (2000). Shopfloor innovation: Facilitating the suggestion and implementation of ideas. *Journal of Occupational and Organizational Psychology, 73*, 265–285.

Backman, M., Börjesson, S., and Setterberg, S. (2007). Working with concepts in the fuzzy front end: Exploring the context for innovation for different types of concepts at Volvo Cars. *R&D Management, 37*(1), 17–28.

Bonner, J. M., Ruekert, R. W., and Walker Jr., O. C. (2002). Upper management control of new brown product development projects and project performance. *Journal of Product Innovation Management, 19*, 233–245.

Bougrain, F., and Haudeville, B. (2002). Innovation, collaboration and SMEs internal research capacities. *Research Policy, 31*(5), 735–747. doi:10.1016/S0048-7333(01)00144-5.

Bouncken, R. B., and Kraus, S. (2013). Innovation in knowledge-intensive industries: The double-edged sword of coopetition. *Journal of Business Research,* 1–11. doi:10.1016/j.jbusres.2013.02.032.

Brentani, U. De, and Kleinschmidt, E. J. (2004). Corporate culture and commitment: Impact on performance of international new product development programs. *Journal of Product Innovation Management, 21*, 309–333.

Burgelman, R. A., and Sayles, L. R. (1988). *Inside Corporate Innovation.* New York, NY: The Free Press.

Büschgens, T., Bausch, A., and Balkin, D. B. (2013). Organizational culture and innovation: A meta-analytic review. *Journal of Product Innovation Management, 30*(4). doi:10.1111/jpim.12021.

Calantone, R. J., Harmancioglu, N., and Droge, C. (2010). Inconclusive innovation "returns": A meta-analysis of research on innovation in new product development. *Journal of Product Innovation Management, 27*(7), 1065–1081. doi:10.1111/j.1540-5885.2010.00771.x.

Carayannis, E., and Alexander, J. (1999). Winning by co-opeting in strategic government–university–industry R&D partnerships: The power of complex, dynamic knowledge networks. *Journal of Technology Transfer, 24*(2/3), 197–210. Retrieved from http://link.springer.com/article/10.1023/A%3A1007855422405.

Chandy, R. K., and Tellis, G. J. (1998). Organizing for radical product innovation: The overlooked role of willingness to cannibalize. *Journal of Marketing Research, 35*, 474–487.

Chandy, R. K., and Tellis, G. J. (2000). The incumbent's curse? Incumbency, size, and radical product innovation. *Journal of Marketing, 64*, 1–17.

Comanor, W. S., and Scherer, F. M. (2013). Mergers and innovation in the pharmaceutical industry. *Journal of Health Economics, 32*(1), 106–113. doi:10.1016/j.jhealeco.2012.09.006.

Cooper, R. (2011). *Winning at New Products.* Cambridge, MA: Perseus Publishing.

Cumming, B. (1998). Innovation overview and future challenges. *European Journal of Innovation Management, 1*, 21–29.

Damanpour, F. (1991). Organizational innovation: A meta-analysis of effects of determinants and moderators. *Academy of Management Journal, 34*, 555–590.

Damanpour, F., Szabat, K. A., and William, M. E. (1989). The relationship between types of innovation and organizational performance. *Journal of Management Studies, 26*, 587–601.

Denison, D. R. (1996). What is the difference between organizational culture and organizational climate? A native's point of view on a decade of paradigm wars. *Academy of Management Review, 21*(3), 619–654.

Dorabjee, S., Lumley, C. E., and Cartwright, S. (1998). Culture, innovation and successful development of new medicines: An exploratory study of the pharmaceutical industry. *Leadership & Organization Development Journal, 19*(4), 199–210.

Droge, C., Calantone, R., and Harmancioglu, N. (2008). New product success: Is it really controllable by managers in highly turbulent environments? *Journal of Product Innovation Management, 25*(3), 272–286. doi:10.1111/j.1540-5885.2008.00300.x

Drucker, P. (1999). *The Frontiers of Management: Where Tomorrow's Decisions Are Being Shaped Today.* New York, NY: Penguin Books.

Dubois, A., and Fredriksson, P. (2008). Cooperating and competing in supply networks: Making sense of a triadic sourcing strategy. *Journal of Purchasing and Supply Management, 14*(3), 170–179.

Faems, D., Van Looy, B., and Debackere, K. (2005). Interorganizational collaboration and innovation: Toward a portfolio approach. *Journal of Product Innovation Management*, 22, 238–250.

Gnyawali, D. R., and Park, B. R. (2009). Coopetition and technological innovation in small and medium-sized enterprises: A multilevel conceptual model. *Journal of Small Business Management*, 47(3), 308–330.

Gopalakrishnan, S., and Bierly, P. E. (2006). The impact of firm size and age on knowledge strategies during product development: A study of the drug delivery industry. *IEEE Transactions on Engineering Management*, 53(1), 3–16. doi:10.1109/TEM.2005.861807.

Guellec, D. (1999). Organizational innovation and organizational change. *Annual Review of Sociology*, 25, 597–622.

Han, J. K., Kim, N., and Srivastava, R. K. (1998). Market orientation and organizational performance: Is innovation a missing link? *Journal of Marketing*, 62(4), 30. doi:10.2307/1252285

Hotz-Hart, B. (2000). Innovation networks, regions and globalization. In G. L. Clark, M. P. Feldman, and M. S. Gertler (eds.), *The Oxford Handbook of Economic Geography*. Oxford: Oxford University Press.

Hurley, R. F., and Hult, G. T. M. (1998). Innovation, market orientation, and organizational learning: An integration and empirical examination. *Journal of Marketing*, 62(3), 42. doi:10.2307/1251742

Jalles, J. T. (2010). How to measure innovation? New evidence of the technology–growth linkage. *Research in Economics*, 64(2), 81–96. doi:10.1016/j.rie.2009.10.007.

Kanter, R. M. (2006). Innovation: The classic traps. *Harvard Business Review*, November.

Kanter, R. M. (2010). Block-by-blockbuster innovation. *Harvard Business Review*, 88(5), 38–38.

Knox, S. (2002). The boardroom agenda: developing the innovative organisation. *Corporate Governance*, 2(1), 27–36.

Koen, P., Ajamian, G., Burkart, R., Clamen, A., Davidson, J., Amore, R. D., Elkins, C., et al. (2001). Providing clarity and a common language to the "fuzzy front end." *Research Technology Management*, 44, 46–55.

Kotler, P. (1991). *Marketing Management: Analysis, Planning, Implementation, and Control*. Englewood Cliffs, NJ: Prentice-Hall.

LeRoy, P., and Yami, S. (2009). Managing strategic innovation through coopetition. *International Journal of Entrepreneurship and Small Business*, 8(1), 61–73.

Lyons, R. K., Chatman, J. A., and Joyce, C. K. (2007). Innovation in services: Corporate culture and investment banking. *California Management Review*, 50(1), 174–192.

Martins, E., and Martins, N. (2002). An organisational culture model to promote creativity and innovation. *SA Journal of Industrial Psychology*, 28(4), 58–65.

McLean, L. D. (2005). Organizational culture's influence on creativity and innovation: A review of the literature and implications for human resource development. *Advances in Developing Human Resources*, 7(2), 226–246. doi:10.1177/1523422305274528.

Medina, C. C., Lavado, A. C., and Cabrera, R. V. (2005). Characteristics of innovative companies: A case study of companies in different sectors. *Creativity and Innovation Management*, 14(3), 272–287. doi:10.1111/j.1467-8691.2005.00343.x.

Menguc, B., and Auh, S. (2010). Development and return on execution of product innovation capabilities: The role of organizational structure. *Industrial Marketing Management, 39*(5), 820–831.

Miron, E., Erez, M., and Naveh, E. (2004). Do personal characteristics and cultural values that promote innovation, quality, and efficiency compete or complement each other? *Journal of Organizational Behavior, 25,* 175–199. doi:10.1002/job.237.

Nieto, M. J., and Santamaria, L. (2007). The importance of diverse collaborative networks for the novelty of product innovation. *Technovation, 27*(6-7), 367–377.

Olausson, D., and Berggren, C. (2010). Managing uncertain, complex product development in high-tech firms: In search of controlled flexibility. *R&D Management, 40*(4), 383–399.

Olson, E. M., Slater, S. F., and Hult, G. T. M. (2005). The performance implications of fit among business strategy, marketing organization structure, and strategic behavior. *Journal of Marketing, 69,* 49–65.

Plotnikova, T. (2010). Success in pharmaceutical research: The changing role of scale and scope economies, spillovers and competition. Jena Economic Research Papers.

Rosenbusch, N., Brinckmann, J., and Bausch, A. (2011). Is innovation always beneficial? A meta-analysis of the relationship between innovation and performance in SMEs. *Journal of Business Venturing, 26*(4), 441–457. doi:10.1016/j.jbusvent.2009.12.002

Sorescu, A. B., Chandy, R. K., and Prabhu, J. C. (2003). Sources and financial consequences of radical innovation: Insights from pharmaceuticals. *Journal of Marketing, 67,* 82–102.

Tatikonda, M. V. (1999). An empirical study of platform and derivative product development projects. *Journal of Product Innovation Management, 16,* 3–26.

Tatikonda, M. V, and Rosenthal, S. R. (2000). Successful execution of product development projects: Balancing firmness and flexibility in the innovation process. *Journal of Operations Management, 18*(4), 401–425. doi:10.1016/S0272-6963(00)00028-0.

Tellis, G. J., Prabhu, J. C., and Chandy, R. K. (2007). Measuring the culture of innovation. *MIT Sloan Management Review, 48*(4).

Tellis, G. J., Prabhu, J. C., and Chandy, R. K. (2009). Radical innovation across nations: The preeminence of corporate culture. *Journal of Marketing, 73*(1), 3–23. doi:10.1509/jmkg.73.1.3.

Tollin, K. (2008). Mindsets in marketing for product innovation: An explorative analysis of chief marketing executives' ideas and beliefs about how to increase their firms' innovation capability. *Journal of Strategic Marketing, 16*(5), 363–390. doi:10.1080/09652540802481934.

Tuominen, M., Rajala, A., and Möller, K. (2004). How does adaptability drive firm innovativeness? *Journal of Business Research, 57,* 495–506.

U.S. Food and Drug Administration. (2012). FY 2012 innovative drug approvals: Bringing life-saving drugs to patients quickly and efficiently. Retrieved from http://www.fda.gov/AboutFDA/ReportsManualsForms/Reports/ucm276385.htm.

Van de Ven, A. H. (1998). Central problems in the management of innovation. *Management Science, 32*(5), 590–607.

Van de Ven, A. H., and Angle, H. L. (1989). An introduction to the Minnesota innovation research program. In A. H. Van de Ven, H. L. Angle, and M. S. Poole (eds.), *Research on the Management of Innovation* (pp. 3–30). New York, NY: Harper and Row.

Vitols, S. (2002). Shareholder value, management culture and production regimes in the transformation of the german chemical-pharmaceutical industry. *Competition and Change, 6*(3), 309–325.

West, M. A., Smith, H., Feng, W. L., and Lawthom, R. (1998). Research excellence and departmental climate in British universities. *Journal of Occupational and Organizational Psychology, 71*, 261–281.

Williams, M. A., Kochhar, A. K., and Tennant, C. (2007). An object-oriented reference model of the fuzzy front end of the new product introduction process. *International Journal of Advanced Manufacturing Technology, 34*(7-8), 826–841. Retrieved from http://link.springer.com/article/10.1007/s00170-006-0645-9.

Wolfe, R. (1994). Organizational innovation: Review, critique and suggested research directions. *Journal of Management Studies, 31*, 405–431.

chapter seven

Moving toward personalized medicine: How to transform product development

Michele Pontinen, Jian Wang, and Christopher Bouton

Contents

What is personalized medicine?

Personalized medicine as an ideal of treating each patient according to his or her personal characteristics (including but not limited to molecular characteristics) is far from reality, if it ever is to become reality. The paradigm shift toward such an ideal, however, is clearly under way. Evidence is pointing to personalized medicine as an ongoing and irreversible trend. What then, are the scientific activities that are performed today that will push us as an industry and as a society toward the future of personalized medicine?

Personalized medicine in its truest sense may never be achievable but can be increasingly approximated. The first step is in the form of molecularly targeted therapies with companion diagnostics (such as Herceptin and Zelboraf). These therapies only work for a certain percentage of the patient population that has a specific genetic signature. As disease biology is better understood, more fine-grained molecular signatures and corresponding single or "cocktail" therapies will increasingly shrink the size of the suitable population toward the direction of "personalization." Targeted therapies require validation in clinical trials with biomarkers to segment patient populations. Such patient segmentation strategy is enabled by in-depth understanding of the disease pathways involved, which requires translational research studies based on well-annotated clinical samples properly managed in biobanks to ensure quality of the samples as well as consented use for research purposes. What is described here are all activities that happen every day in pharmaceutical and clinical research. Personalized medicine is not an empty dream; it is a destination and we are on our way, making progress every day.

There are several drivers propelling us along the path to personalized medicine (Figure 7.1):

1. Increased knowledge of normal and disease biology on the molecular level, accelerated by large government-sponsored initiatives, such as the Human Genome Project, is making personalization technically possible and ethically essential.
2. Failure of the "one-size-fits-all" blockbuster model of the old pharmaceutical business model.
3. Comparative effectiveness and value-based costing from the payer's perspective.

There is strong scientific underpinning for the targeted therapy approach. Today's targeted therapies are outcomes from the translational research paradigm that connects molecular understanding of human health with the clinical care aspects. As shown in Figure 7.2, the field of translational research (first coined by a PubMed-indexed paper in 1993) has grown exponentially as judged by number of related scientific publications. The federal government has established funding mechanisms such as the National Institutes of Health's (NIH) Clinical and Translational Science Award (CTSA) to support this approach. Pharmaceutical companies have also embraced this new paradigm, especially in therapeutic areas where the molecular mechanisms of diseases are relatively well known, such as oncology and immunology. Over the past several years, the pharmaceutical industry has evolved itself to align with this new scientific paradigm to put much more emphasis on understanding biological mechanisms in a "learn and confirm" model of drug discovery and

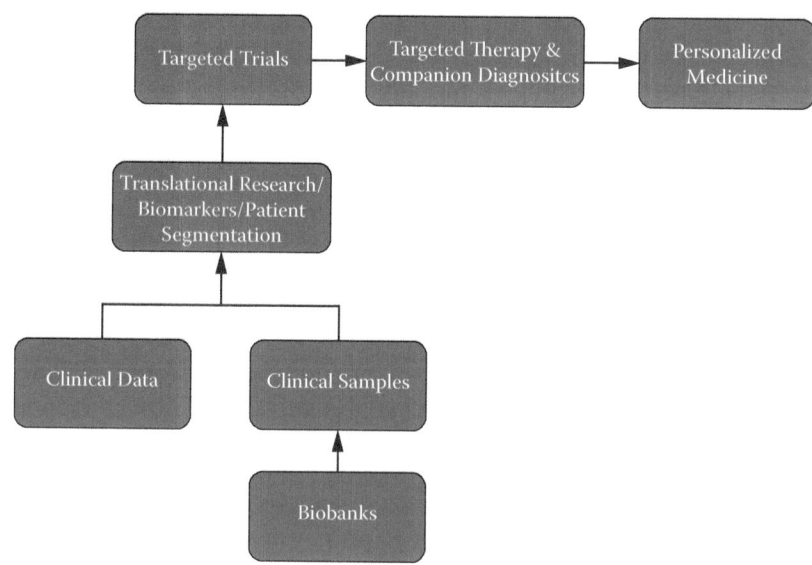

Figure 7.1 Getting to personalized medicine—the big picture.

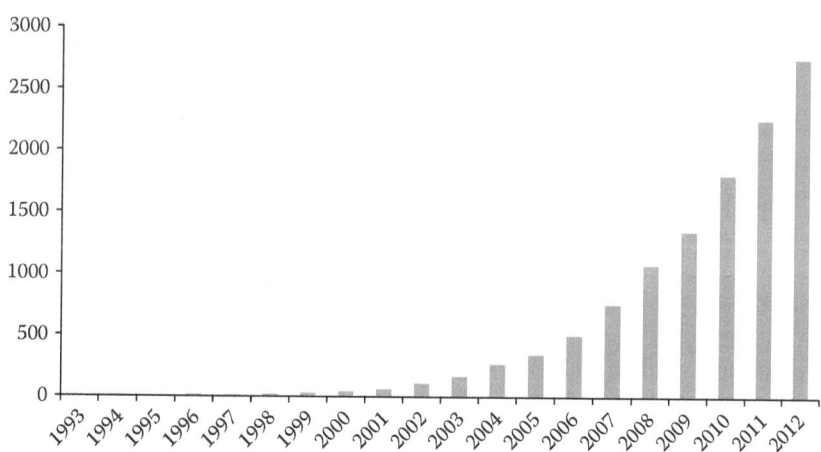

Figure 7.2 References to translational research, translational medicine, or personalized medicine. Number of publications in PubMed per this formula: PubMed search phrase: ("translational research" or "translational medicine" or "personalized medicine") AND **** [Publication Date], where "****" is to be replaced by the year in question.

development. Recent successes have shown remarkable validation of this approach. For example, Vemurafenib (Zelboraf), which specifically targets the melanoma patient population with BRAF V600E mutation, demonstrated over 50% response rate, compared to the 10% to 20% response rate of the prior approved standard therapy (Fleming et al. 2011).

Targeted therapy as a business model is also showing traction. If nothing else, the alternative is not an option, as clearly stated by John Lechleiter, CEO (then COO) of Eli Lilly: "In my industry, we would be powerless to resist personalized medicine, not to say foolish" (Parsons 2007). There are several considerations that make targeted therapy an attractive business model:

1. A more targeted approach has a higher prospect of large therapeutic effects and fewer safety concerns within the chosen population, likely leading to smaller trial size and shorter and potentially cheaper trials, notwithstanding the additional complexity that is introduced by companion diagnostics. The Food and Drug Administration's (FDA) new "Breakthrough Therapy" designation reflects, and indeed validates, this approach (McCaughan 2012).
2. Shorter trials lead to longer patent protection of marketed products.
3. Better measured efficacy and safety profiles lead to a higher likelihood and level of reimbursement from payers.
4. Biology is a linked network. What is a targeted therapy for one indication can often be expanded into other indications with the same or related underlining biological mechanisms, either individually or as part of a combination therapy approach. There are many examples of drugs first approved in a small niche indication that grow into much more commercially successful territories upon additional approvals, Gleevec being a good case study ("Gleevc" 2012).

Today, it is often not enough to demonstrate a medicine's safety, efficacy, and quality for it to be a commercial success. A "fourth hurdle," reimbursement, must also be overcome. In order to reduce healthcare costs, governments and payers are becoming increasingly sophisticated and insistent on evaluating a medicine's clinical effectiveness as well as cost effectiveness to answer questions in the realm of if the new product is better than the existing alternatives and does it provide good value for the money that the pharmaceutical company is charging. Even physicians are questioning the cost of drugs and treatments, as evidenced by a recent article on "unsustainable cancer drug prices," a perspective by CLM experts on the price of drugs (Experts in Chronic Myeloid Leukemia 2013). The personalized medicine approach will make it easier to overcome this hurdle.

Impact of translational research and personalized medicine on the pharmaceutical development process

As discussed, the future of personalized medicine is reflected in today's translational research and biomarker- and molecular-diagnostics-enabled targeted therapy. With the genomics revolution now well underway, the line between translational research and early stage clinical development has started to blur. Technologies that have been developed over the last few decades for the discovery and analysis of genetic sequence information can now be effectively applied in the clinical development process. In particular, genome sequencing technologies developed during the Human Genome Project (HGP) have allowed early basic researchers to elucidate the structure of genomic information and the function of genes in cellular systems. These insights into the machinery of life have been foundational and critical for the understanding of gene functions both in normal and disease systems. Technological advances since the HGP have commoditized genome sequencing and other related technologies. For example, it took roughly a decade and $3,000,000,000 dollars to sequence the first human genome in the 1990s. It is predicted that within one or two years, sequencing a human genome will be reduced to one day and about $1,000 (Wetterstrand n.d.). Such commoditization of molecular platform technologies is allowing for the migration of these same approaches into the clinical development realm.

An excellent example of this migration of approaches from translational research to clinical development is the recognition of the importance of cancer's genetic makeup, not just its organ of origin in the body, for the progression and etiology of the disease. The classic clinical notion of tissue dependence for many diseases does not play out in the same way with cancer. Instead, it matters not as much whether a cancer is in the intestine or brain as it does whether it has a specific genetic alteration such as p53, HER2, or AKT1 mutation. What this means for early clinical development is that the most important thing for a clinician to know may not be whether a cancer is in a specific organ but what gene mutations have enabled its formation. With that as a driver, clinical development for the sake of finding the right patient cohorts and treatment design relies upon the same technologies that are used to elucidate the basic cellular functions of genes such as sequencing, gene expression detection, and bioinformatics approaches. This molecular-mechanism-based approach is also enabling industry to revisit candidates dropped from further development because of unacceptable safety and efficacy findings. Several companies have begun to adopt a systematic approach to repurpose failed candidates. Drug rescue and repurposing based on understanding of molecular mechanisms is proving successful. Many believe, for example, without a biomarker-based patient segmentation strategy, Herceptin would never have been approved, let alone becoming a billion dollar drug. As scientific knowledge of the disease-state drives treatment approaches to focus on the molecular mechanism, perhaps a more affordable and faster path to successful development can become more viable for both new and earlier failed drug candidates. The extreme case of "rescuing" a drug is when the use of the medication is limited after its approval, based on new understanding of the molecular mechanism of the drug. The FDA approval history of Erbitux (Cetuximab) is a good case study (Metcalfe 2009).

The progression from translational research to clinical development should not come as a surprise. Once the HGP generated enough data to provide the basic map of human genetics, it was only a matter of time before the research tools, used to investigate biological systems at the basic research level, matured enough to be applied in the clinical realm. Furthermore, the clinical realm itself is a relatively new discipline and hence has a driving need to further integrate approaches from early basic research in order to advance. The concept of applying specific, small molecule compounds for the remediation of specific ailments is a very recent occurrence in the history of humanity. It was only 1897 when Aspirin, the first "blockbuster" synthetic drug, was first purified by Felix Hoffmann, a chemist with the German company Bayer, based on knowledge of the therapeutic benefits of willow bark. What we know of clinical utilization of therapeutics and the development of such therapeutics largely evolved since that time. Since then, the vast majority of "modern" pharmaceutical therapeutics have been developed on the false premise that all humans are similar enough to derive roughly the same therapeutic benefit from a molecule. Outside of a handful of notable exceptions, we know that this is untrue. The genetic background of humanity is diverse enough that the same molecule in different individuals can produce radically varying results ranging from toxicity to efficacy. The development of therapeutics, without regard to this biological fact for the past century plus, was, in large part, due to our inability to differentiate patients based on genetic backgrounds and effective biomarkers. This led to massive inefficiencies in the design and clinical development of therapeutics. Modern translational research approaches though, when applied in the pharmaceutical development process, mitigate these issues, increasing successful treatment strategies and broadening options for application.

Moving forward, it seems natural that the blurring of the lines between translational research and clinical development will continue for a long time to come. Genes and their mutations are only the beginning of what is a complex interplay between genes, environmental context, and systems biology. As more and better approaches are developed in the research space for the study of these aspects of biological systems, their commoditization, through performance and efficacy increases, will allow for application in the clinical space to enable even finer-grained application of therapeutics based on an understanding of a specific patient's genetic makeup and disease parameters.

An important display of the impact of translational research and personalized medicine on the pharmaceutical development process is the increasing importance of access to human specimens. All the biomarker-based work described earlier is dependent upon human samples from both disease and normal physiological states. Collection and biobanking of human tissues and other samples for the purpose of long-term research use, which until recently had primarily been the domain of academic research hospitals, is becoming an integral component of the pharmaceutical (including devices) research and development (R&D) process and a critical enabling technology. Human samples are the best approximation of the human subject for personalized medicine research. As such, it is vitally important to properly handle the samples from initial acquisition to final disposal according to rigorous scientific and operational standards. In addition, for the scientific community to enjoy the privilege of access to human samples, the utmost care must be applied to respect patient informed consent. Many institutions (both hospitals and pharmaceutical industry) are beginning to routinely consent patients for broad, long-term "future use" samples, increasing the complexity of managing such consents, sometimes over a period of tens of years.

One of the key challenges in clinical trial and "future use" biobanking is effectively managing the staggering complexity of sample collection, consent, storage, and destruction of samples obtained in the trial and their future use. A typical complex workflow is illustrated in Figure 7.3:

1. For each clinical trial, the setup information and biospecimen processing logistics are sent from the sponsor to the external biorepository and contract research organization (CRO) partner.
2. The types and specifications of biospecimens to be collected (and collection kits) are sent from the external biorepository and CRO partner to the trial sites (hospitals, clinics, and other healthcare organizations) either directly or via a central laboratory.
3. At the trial sites, subject consent and clinical information are obtained and typically recorded in an electronic data collection (EDC) system.
4. The collected biospecimens are sent from the trial sites (along with certain clinical information about the patient) to be processed or stored at the external biorepository partner location.

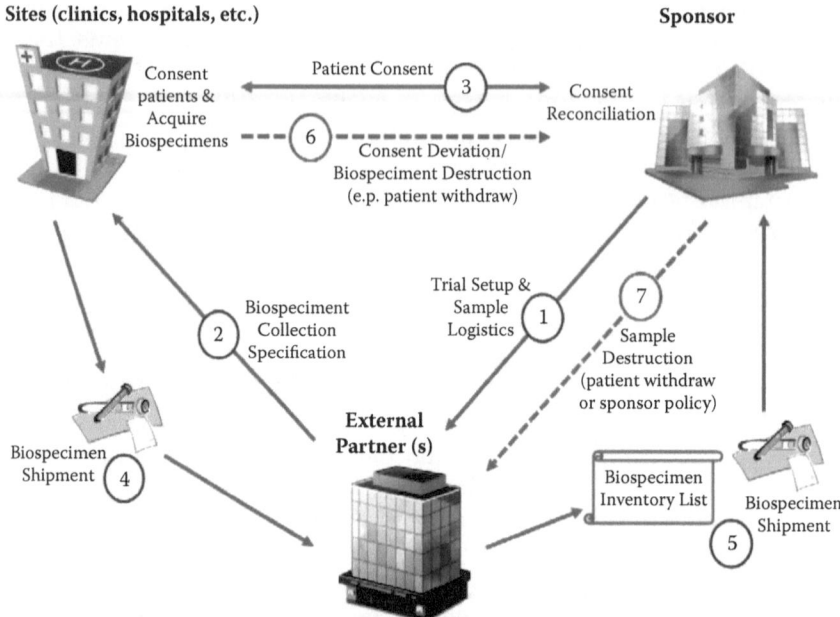

Figure 7.3 Sample collection workflow.

5. On a regular basis, biospecimen inventories are sent from external biorepository partners to the sponsor, which performs specimen reconciliation and inventory update.
6. The sponsor tracks specimen permissions and consents (e.g., site specific storage durations, patient withdrawal of consent).
7. Biospecimen destruction requests are sent from the sponsor to the external biorepository partner due to consent or study modifications, or for inventory management.

Changing relationships among biopharmaceutical industry, healthcare providers, payers, government, advocacy groups, and patients

Personalized medicine is a paradigm shift that will have the potential to impact the healthcare ecosystem in profound ways. One of the most visible aspects is the evolving drug development model that is significantly reshaping relationships among all healthcare and industry stakeholders—the biopharmaceutical, device and diagnostics industry, regulatory agencies, research centers, those delivering care, payers reimbursing the cost of services and products as well as physicians, patient advocacy groups, and indeed individual patients themselves. It now appears that to successfully compete in this new environment, every healthcare stakeholder will need to change the way they work and the way they work together. They will need to collaborate in order to survive and thrive. Collaboration means contributing and sharing the risk and responsibility of product development. This new way of working will take shape as the "web of stakeholders" bring not only their expertise but also their data to the table. Together, they will collaborate; influence; and make the who, what, when, where, and how decisions for candidate development. We see more than just a change in relationship; we see the opportunity to transform how and what products and treatments are developed and their subsequent commercialization.

Transparency among the stakeholders in this model will be critical, and the open exchange of data (with privacy concerns addressed) will evolve as the norm. This will also change how we think about product development—especially about activities and decisions that may no longer be made within our own organizations. Data-driven decision making will become the prerequisite for this new product codevelopment model. Relationships among stakeholders will be strategic, and critical new competencies will be required to negate any concerns on trust and transparency, as well as risk and profit sharing. This new way of working will need to deliver tangible financial and economic benefits to all stakeholders if it is to be successful. Buy-in from regulators will be critical and how we measure quality and the benefit of care or services, products, or treatments will change. We evidence this change today as government, consumers, healthcare payers, physicians, and providers are requiring proof of safety, quality, value, and price. Today, innovations in science and information technology can move this change forward at a rapid pace.

Leveraging "big data" captured and owned by healthcare partners

A trend that intersects personalized medicine and the new healthcare stakeholder ecosystem is "big data." Healthcare stakeholders collect vast amounts and variety of data at an increasing speed, making pharmaceutical product development a prime candidate to benefit from big data technologies. However, leveraging such data, now considered extremely valuable and proprietary, will be a challenge for some time. Our industry (biopharmaceutical, device, and diagnostics) is awash in data, information, and knowledge. These data are for the most part in disparate formats, on multiple platforms, often in silos, with minimal data integration. Big healthcare players (payers and providers) have launched successful, independent EHR (electronic health record) initiatives. Individual patients are also starting to maintain their own personal health records (PHRs). Several large research institutes (hospitals), patient care centers, and payers are beginning or have experienced some success in constructing their own health information architecture models that support internal data integration and data mining. Some have started data exchanges that support real-time data sharing with external partners or collaborative ventures. The biopharmaceutical, device, and diagnostic industry is realizing the importance and value of data to help make more informed decisions, and today we are beginning to see a new understanding and appreciation for external data sharing and exchanges. Industry is now coming to appreciate that external data-driven collaboration can transform their business, making them more competitive and success more predictable. We have witnessed some big-data exchange stories, but these are generally realized through database subscription sales and not formal partnerships where risk, scientific discovery, and profit are components of the relationship model.

Experience is today telling us that the "data owner" will most probably drive the construct of the exchange and this new lever will change the traditional product development model. Collaborative relationships among the healthcare entities will determine data or information fit for use as well as integration requirements. This will most likely occur once data standardization (interoperability and industry data standards) moves to a more mature state. As data exchange volumes increase and more stakeholders become participants in the new model, poor quality data will become an increasingly complex problem to address (Bertolucci 2013). Most agree that quality data is critical when making a medical decision about a subject in a clinical trial or a patient receiving treatment from their healthcare provider. This convention will be no less important when making the decision concerning product development. As economics enter the relationships, transparency will be crucial. The new model, based on transparency (trust amongst the parties), also requires the assurance that data is trustworthy (veracity and authenticity). This "data-quality lever" should instantiate governance where data-driven decisions are made to identify, select, and optimize product development candidates that are fit for a specific, segmented population.

Evolving relationship between healthcare providers and industry

Traditionally, the relationship between industry, healthcare providers, and regulators is focused on the clinical trial process and postapproval studies (Figure 7.4). This model is no longer sufficient in the personalized medicine era. Their relationship and interaction needs to be much more intimate, starting long before the first patient enters a clinical trial and lasting long after any mandated postapproval studies. For example, with the targeted therapy approach, industry will need to have access to patient medical information and biospecimens to connect clinical and

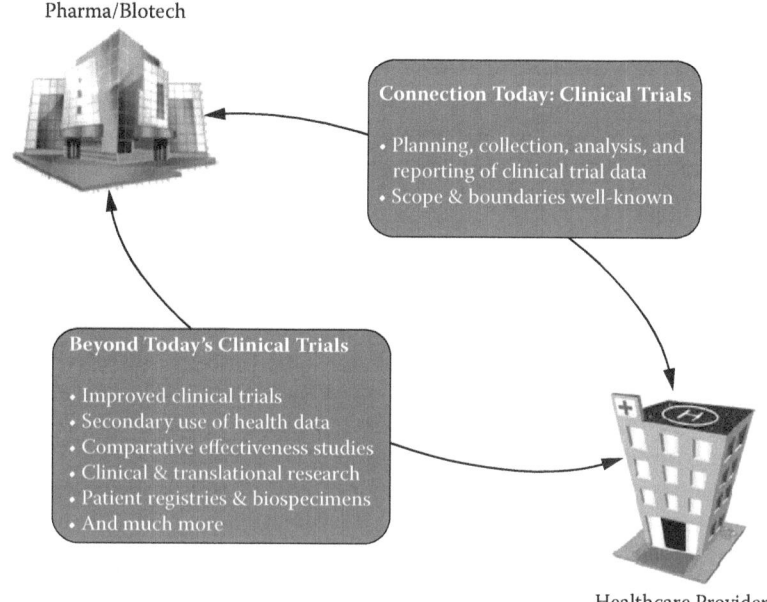

Pharma/Blotech

Connection Today: Clinical Trials
• Planning, collection, analysis, and reporting of clinical trial data
• Scope & boundaries well-known

Beyond Today's Clinical Trials
• Improved clinical trials
• Secondary use of health data
• Comparative effectiveness studies
• Clinical & translational research
• Patient registries & biospecimens
• And much more

Healthcare Provider

Figure 7.4 Relationship between HC providers and industry.

molecular profiles in the very early stages of its R&D process. This need is also very relevant postapproval, as real-world evidence from the health-care setting (as opposed to controlled clinical trials) is a strong influencer on payer reimbursement decisions. As such, the ability to construct a collaboration model where clinical information and patient samples can be used to support personalized medicine products is a winning strategy for all parties involved in the healthcare ecosystem. As noted earlier and by others (Green and Troyanskaya 2012), data-driven functional genomics strategies are combining statistical methods and computer science to integrate diverse experimental information and data for the purpose of making novel, biological predictions. This "biological relevance" is not currently well characterized in the type of traditional clinical trials we conduct today (Huttenhower et al. 2006). As these new knowledge-based opportunities present themselves, healthcare providers and industry will see the value of contributing and shaping a new development model with their scientific expertise, business acumen, and their data. The new model will leverage both internal and external scientific expertise, combined with data-driven decisions, to predict and deliver safe treatments, disease prevention, and perhaps cure. This new model can help realize targeted therapies for molecularly defined patient segmentation.

A robust electronic health record (EHR) system is a tool that cannot only help to improve healthcare and provide critical information for ethical product development, but also significantly move the process forward to realizing targeted therapies. EHRs are able to access vast amounts of patient data and information about medical history, disease, and prior treatments. Today's EHRs are not yet perfect for personalized medicine, but their shortcomings are largely understood and can be addressed over time. Classes of data available for mining and analysis via EHRs (as well as the associated challenges) include: billing data (consisting of codes); laboratory and vital sign information (longitudinal records); provider documentation (data and information generally required for all billing); documentation from reports and tests (a mixture of structured and unstructured data where often the specific report or test result is not easily available); and medication records (which vary in format and structure within EHRs). Depending on the purpose of the EHR system, structures vary, multiple formats (or data classes) exist, and quality is sometimes questionable; making the transformation from data (e-record) to information to knowledge sharing challenging.

Unlike countries with established universal healthcare, it is challenging within the United States to begin the wholesale, stepwise creation and standardization of an EHR system, with agreed-upon universal security requirements. Today's EHR weaknesses include: availability of data type, ability for recall (to query), and precision (variability). ICD (International Classification of Diseases) and CPT (Current Procedural Terminology) codes are in common use, yet recall and precision is variable and data often missing. Lab data, although precise and mostly structured, may need to be aggregated differently, as variations of the same data element often occur and the value is dependent on the test. Also, with all lab findings, normal ranges and units change over time and will impact interoperability. Medical records and clinical documentation vary as well impacting precision and ability to process (Denny 2012). The weaknesses inherent with several EHR systems are being addressed with the adoption of standards (i.e., LOINC® [Logical Observation Identifiers Names and Codes]), a Consolidated Health Informatics standard for lab test names, also part of HL7; CPOE (computerized provider order entry) for managing hospitals stays and inpatient medication records mapped to controlled vocabularies; DICOM (Digital Imaging and Communications in Medicine). a standard for transmitting medical imagine; CDISC (the Clinical Data Interchange Standards Consortium) data standards for clinical trials; and HL7 standards (e.g., for discharge summaries, summary patient records). Much work still remains before we witness a robust return on investment (ROI) for EHR for research purposes.

The most comprehensive effort to establish a standard certified EHR in a secure environment and within a nationwide health information network is the Federal Health IT Community. This community is making funds available to participants that actualize EHRs within the United States. Federal Heath IT is operationalizing the effort under the Health Information Technology (HIT) initiatives. Funding for this program was originally provided by the American Recovery and Reinvestment Act of 2009, authorizing the Centers for Medicare & Medicaid Services (CMS) to award incentive payments to those parties who would demonstrate a "meaningful use" of a certified EHR. Thirty-two federal agencies (and 2,000 additional organizations both in government and the private sector) are currently participating in the Federal Health Information Model (FHIM). Financial penalties are scheduled to take effect in 2015, for Medicare and Medicaid providers who do not transition to EHRs ("HIE Market Report" 2011). To date, federal agencies, such as the Department of Defense and the Department of Veterans Affairs, are beginning to actualize the value of HIT (Byrne et al. 2010). For those physicians in the process of adopting EHR systems, a survey released by Health and Human Services (HHS) in July 2012, indicated that 55% of those responding said they adopted at least some EHR technology in their practices, and 85% of physicians who adopted EHRs said they were somewhat (47%) or very (38%) satisfied with their systems. A majority of the physicians said they would purchase their EHR systems again, further indicating their satisfaction with the new capability technology (Jamoom et al. 2012).

Moving a step forward, as a result of a retrospective review, HHS is now developing a final rule that would make significant modifications to the Health Insurance Portability and Accountability Act (HIPAA) Privacy, Security, Enforcement, and Breach Notification Rules (HHS 2012). This rule change could streamline the process for researchers and scientists to obtain HIPAA authorizations for their research; it may also prove to harmonize privacy rule procedures with informed consent requirements incorporated into the Common Rule.

EHRs (actualized under the HIT initiatives) have grown significantly since 2001, negating the argument that physicians will not invest. It appears physicians too are realizing how this capability impacts their bottom line. Office-based physician use increased from 18% in 2001 to 72% in 2012 (Hsiao and Hing 2012). Although not as advanced as EHR initiatives in the United Kingdom and by other countries, U.S. government incentives, coupled with rules impacting reimbursement, are driving up the rate of EHR adoption.

Recently, DNA bio-banks are becoming associated with EHR systems, driven by the personalized medicine vision to enable EHR-based genomic science. This is occurring at Marshfield Clinical (WI), Northwestern University's NUgene Project, and Vanderbilt University's BioVU as well as Kaiser Permanente (which has genotyped >100,000 individuals) ("Kaiser Permanente" 2011). As more institutions and healthcare providers and payers capture biomarker data and deploy their EHR platforms, the relationship among the major groups begins to level. This leveling effect will present more challenges but also great opportunities for all healthcare stakeholders.

Influence of payers

As noted, industry must no longer just clear the hurdle of regulatory approval to sales. Now they must also clear the hurdle to reimbursement. Payers, including the federal government, are using their leverage to set reimbursement for products and treatments. This is clearly impacting industry's strategy for product development as well as their expectations for profit. Today's influence by payers, added to late-stage failure, diminishing pipelines and patent expiratory are significantly impacting the process and the bottom line. This "external influence group" now must be brought directly into the product development process.

Access to and mining of payer and provider data will drive a new relationship as well as influence product development. Tapping into their intellectual property (their data), the definition of and requirements for the treatment population segmentation, as well as the safety and efficacy targets can be determined and affirmed early in the development lifecycle. Payer influence and input can remove (if the selected candidate is successful) the uncertainty of reimbursement. Payer data is a mixture of structured and free-text clinical information. In light of healthcare reform and the current recession, many payers are beginning to move swiftly forward with data mining efforts to leverage their intellectual capital beyond reporting on financials. The outcomes of partnering with pharma, biotech, and or the device industry may lead to new, less expensive delivery solutions for home health and new products for monitoring and diagnosing specific treatment populations—something akin to personalized medicine. Sharing data, risk and ownership of decisions for product development (coupled, of course, with some type of remuneration) may be the optimal new business model for this time.

We also are witnessing many payers setting the stage to drive their own data-driven healthcare decisions. They are moving to adopt advanced analytical capabilities that will allow them to: predict negative trends, both at the single patient level and in the aggregate; become more proactive by identifying the who at risk; conduct modeling and simulation to determine what is the best combination of services and or treatments for the individual patient or a segmented population; and finally, identify what actions and preventive measures they can take to prevent existing or newly identified negative outcomes (i.e., adverse events or treatments or services that are not effective). The hurdles they face, however, remain: (1) data quality (erroneous and missing data); (2) privacy barriers (i.e., HIPAA restrictions especially for data exchanges among payers); (3) unique infrastructure and information (IT) policy inefficiencies; (4) identifying what is relevant and excluding what is not; and (5) data structure or lack of (it is estimated that approximately 80% is nonstructured and 60% of this data is clinical documentation—data, information, and knowledge not easily available for either the treatment of a patient or insight on managing a specific patient population). As a result, the use of data has mainly focused on retrospective reporting of findings. However, the desire is growing to use data (current, historical, and in the aggregate) in real time to drive better clinical decision making, moving payers away from their current volume-based care model. Payer data and influence can drive critical reforms to change the current model into a more attractive and successful product development model.

Tapping into individual patients and patient advocacy groups

Health advocacy organizations (HAOs) are key stakeholders that are bringing increased pressure on industry to create new products, treatments, and cures at a faster pace. We can trace this back to the organization ACT UP (the AIDS Coalition to Unleash Power), an international, direct-action advocacy group working to impact the lives of people with AIDS. This group pressured (forcefully at times) the government and the pharmaceutical industry for better science leading to improved treatment and eventually a cure for the disease. Their efforts helped push the implementation of "Fast Track" adopted by FDA for the expedited review and approval of ethical products. Today, HAOs are beginning to leverage their membership and influence to serve as a primary source for clinical research trial recruitment. In fact, some are early players in the crowdsourcing of clinical research trials. HAOs are also becoming more transparent about funding they receive from industry. This new transparency is now enabling legislatures, regulators and the public to easily "follow the money and evaluate and dispel bias or conflict of interest concerns" with the funding they receive from industry (Rothman et al. 2011).

With the near ubiquitous availability of the Internet in modern societies and the rapid decrease in the cost for DNA sequencing, patients are increasingly taking their personalized healthcare matters into their own hands, thus giving rise to the personal genomics movement. Self-assembled patient communities, sometimes aided by for-profit organizations, are becoming an important player in the personalized healthcare ecosystem. We see this occurring in researcher-organized and patient-organized studies. Members of PatientsLikeMe and 23andMe are currently participating in research studies, by providing self-reported data, answering surveys, and providing consent for genotyping. Patient-organized studies, conducted by Genomera, Althea Health, and DIYgenomics are making genotyping data and blood test results available. These patient-organized studies are using self-tracking devices to capture and self-report data. Devices such as myZen, FitBit, and TelCare are just a few examples. Crowdsourcing is becoming the recruitment method employed.

Crowdsourcing research trials are the natural, next step for online patient social networks and e-communities. It is believed that crowd-sourcing research provides more opportunities for more levels of openness as well as privacy, as the trial participants decide what data to share with whom (Swan 2012). HAOs and other patient communities will play a more significant role in influencing health policy and can become an influential stakeholder in a new codevelopment model.

The questions of regulatory approval, institutional review board (IRB) review and monitoring, as well as subject (patient) consent still needs to be addressed in the new trial crowdsourcing model. Some HAOs and research institutes who champion this type of research point out the "practical impossibility" for a traditional compliance mechanism (informed consent) and are asking participants "to consent and acknowledge the fact." For example, the Harvard Medical School's Personal Genome Project (PGP) is asking participants to agree to a more open stance on privacy and data use: "the data that you provide as part of participating in the PGP program may be used … to identify you as a participant in otherwise confidential genetic research" ("Informed Consent" 2011). Subjects are consenting to participate, which perhaps indicates a trust in the institution, confidence in social media, or perhaps the influence of HAOs in this space. Others are questioning the validity of crowdsourced research trials, calling them nothing more than "self-treatment experiments." No comment has appeared to date from regulators on the trial crowdsourcing model. Nevertheless, the is currently exploring methods to monitor and contribute to social media conversations, specifically about medical devices ("providing meaningful and timely information about products they regulate"), indicating that even the FDA is recognizing and beginning to appreciate the power and influence of social media (FDA 2013a).

The HAO is a more highly concentrated stakeholder, impacting health-care providers, payers, and physicians as well as the product development industry. They, like all other stakeholders, are driven by social and economic demands. They also must respond to their constituent groups. Their primary demand drivers include: population growth rate; increasing life expectancy; advances in medical care and technology; and patients who are demanding more. Their profitability drivers include: obtaining grants and federal funding; effectively managing patient demand; and referring patients to the most cost-effective products (Global Edge 2012). Whatever their role in this new product development model, they should also bear ownership as well as a responsibility and commitment for success.

New operating model

New ways of working: What will this new model look like?

If we accept the premise that today we are awash in data (both internal and external), then the definition of a codeveloper will also include a wider range of stakeholders than traditionally considered. The industry has, for quite some time, engaged in codevelopment arrangements with external partners. Current collaboration arrangement between partners is quite standard. We see these arrangements as codevelopment and licensing whereby the control for development remains mainly with the major partner. However, we see a new model developing whereby multiple stakeholders are becoming full parties to product development, where governance enables their decision making, assuming risk and responsibility, and success is shared among all making each into a "true codeveloper." This new model will include payers and providers, physicians, geneticists, computational scientists, patients, HAOs and researchers who will bring their expertise to the table; understand the patient impact (the contextual understanding), as well as new scientific knowledge at the molecular level to bear on product development and patient segmentation. In addition, these stakeholders will bring to the table their data, where data-driven decisions actually will prevail. These new stakeholders will move into the true role of changing the product or treatment development model.

In Figure 7.5, we postulate a potential cancer product codevelopment model. Stakeholders in this product model include: a trial sponsor or sponsors (from biotech, pharma, device, diagnostic, or a combination of stakeholders); research institutes or specialty scientific cooperatives that provide a better understanding of the molecular signature of the specific disease biology (its specific variation at the molecular level); diagnostic entities that use specific biomarkers to segment the population for the candidate product or treatment; EHR data from providers and payers to identify the "targeted treatment group"; bioanalysis of clinical data by specialty providers; collaborative up-front decision making by pharmacy benefit managers, physicians, and other payers that set efficacy, safety, and acceptable cost parameters; advocacy groups with direct access to those "qualifying subjects" for trial recruitment; conducting the trial in a "virtual environment" that might be the "cloud" where physicians, treating patients with this specific disease, could be recruited as investigators; and regulators eventually being able to review data in real time, transforming the review and approval process and reducing time to market.

This new model can be realized with close and transparent collaboration by all stakeholders involved in the effort—an extremely new way of working for most. This model also significantly challenges the traditional

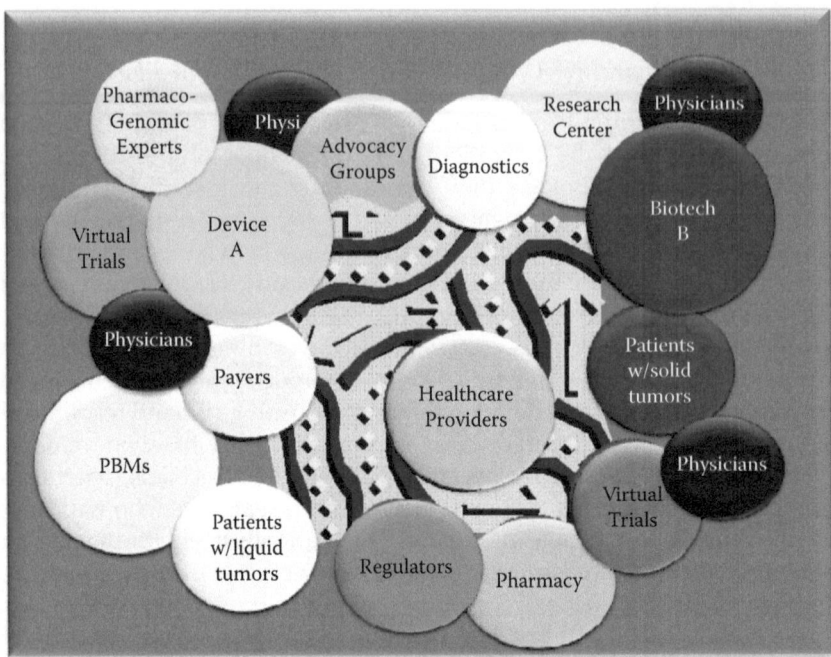

Figure 7.5 Example: Cancer collaborative alliance of codevelopers.

concept of intellectual property. Legal arrangements that deliver value (monetary or other recompense) to each stakeholder and that do not stifle innovation or impair trust will be critical for adoption and success. We have been incrementally moving toward this model for some time now (Novartis n.d.). As industry copes with late-stage failures, weak pipelines, and patent expirations, they must look at other models that increase their chance for success and reasonable reimbursement. Just recently, a major pharmaceutical company entered into an alliance arrangement noted as "one of the largest ever initial payments in a pharmaceutical industry licensing deal that does not involve a drug already being tested in clinical trials" (Moore 2013). As this particular type of model matures, more alliance arrangements will be tested and modified over time. Whatever the alliance construct becomes, a one-size-fits-all will never be the norm. Each party committing to this type of relationship will be shaped and driven by its own need and value drivers and consensus by all involved.

Transparency and the right governance model that assures all stakeholders have an equal seat at the table means they also must accept responsibility for risk. Governance cannot be prescriptive, driven by a single entity. It will need to be fluid and transparent, where decision makers may change along the development path but where all stakeholders take responsibility for the final product or treatment (the safety, quality, and efficacy of the product or treatment and the cost). In addition to clear and present governance, there needs to be other incentives to identify and remediate risk throughout the entire product lifecycle.

How quickly the model is embraced and matures will depend on a multitude of factors:

- The trust factor (especially transparency amongst all stakeholders)
- Buy-in by regulators and other major players (e.g., payers and provider stakeholders) that freely enable and support unbiased input and evaluation
- Responsibility for risk and ownership that is clearly defined, including data ownership, data sharing, standards, and all other conventions that support requirements for evidence; thereby identifying decision makers and what and how their particular share in commercial profits can be well quantified
- Major change management efforts to ensure there is transformation of the current model and that all embrace this new way of working

As we have noted, virtual trials, especially using crowdsourcing enrollment techniques, are starting to become an accepted and adopted practice. This approach to recruiting and conducting trials is truly a paradigm shift for our industry. HAOs were the first to enter this space, enabled by their extensive use and expertise with social media. Recruitment directly by HAOs is proving successful. As noted earlier, FDA is exploring the use of social medial to provide information on devices approved by the agency, recognizing the importance of leveraging this new media. We see crowdsourcing as the next logical extension of the social media paradigm.

Additionally, we see that eventually this new product development model will appear more circular as opposed to flat or linear in shape and process. In the traditional model, candidates move forward through a series of governance boards when key milestones are met. The model generally works when assessments and reviews revisit the original strategy and the decision to move forward is based on current scientific data and knowledge (measured against original assumptions), risk is reevaluated, and public information (i.e., on competitors, in literature, from regulators) is taken into account. The models in Figure 7.6 call out the governance, stakeholders, and the decision-making changes from the traditional to the new model of product development. In this traditional, linear model, the development process generally occurs within the confines of the sponsor company. The overall lifecycle, as shown, is normally governed by the pharma/biotech leadership board, and separate, science, and business focused governance committees manage the candidate as

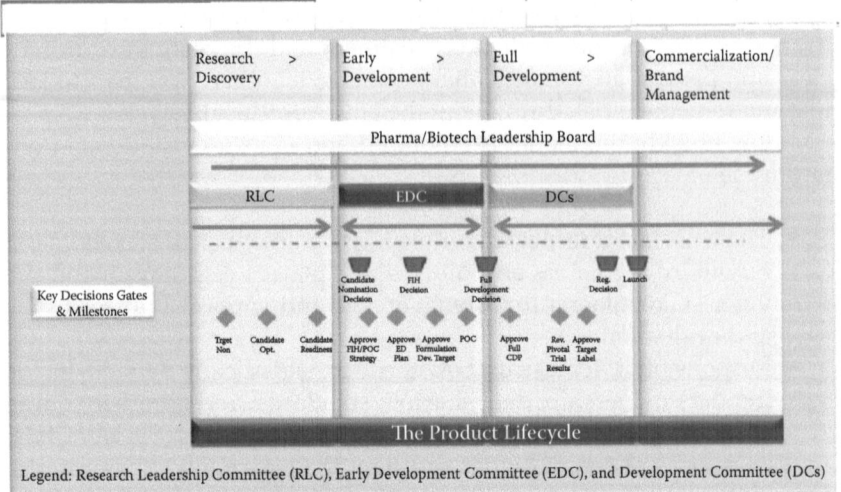

Figure 7.6 Traditional, linear product development model.

it moves through the lifecycle. The candidate is able to move forward as key milestones are met and decisions are made by the respective governance committees. Until recently, mainly internal resources delivered on development services; companies of course conferred with industry experts and research institutes on a limited basis, but internal expertise and decision making carried the day.

The new model, arguably, will appear quite different as more "true" codevelopers become engaged and begin to make decisions for development (Figure 7.7). For success, codevelopers must agree on the several critical conventions:

- The decision to advance a candidate (with a specific target subject segment) is collectively owned by all stakeholders.
- The decision to advance is taken by the codeveloper that is an expert in the space.
- All decisions are data driven (and validated) and clearly transparent.

This new way of working, whereby codevelopers provide expertise and unbiased evidence (data, scientific knowledge, and business acumen) to substantiate or validate a decision, will ensure approving a candidate for advancement (as well as the target subject segmentation) is science based and quantifiable—making the collective knowledge of unconstrained data a reality.

Direct input and decision making by these codevelopers may well continue even after approval and commercialization of the product. The

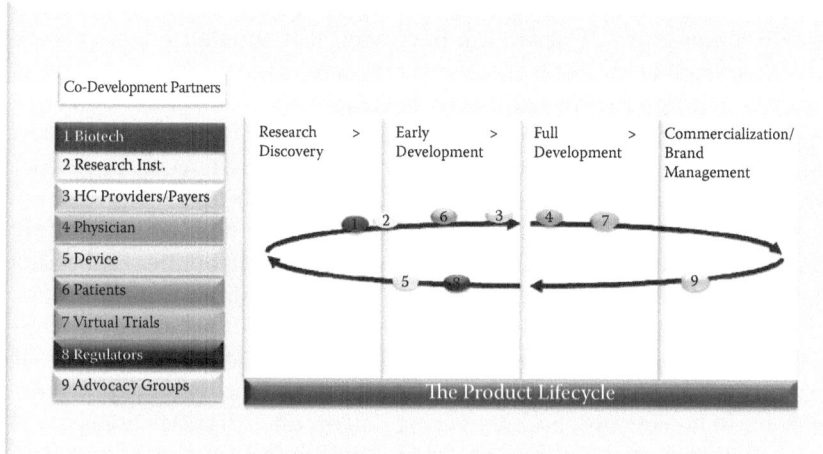

Figure 7.7 A "new look" for the product development model.

greatest promise for this type of model is the ability to tap into and leverage the expertise of codevelopers that enables making data-driven decisions (such as "kill early" before additional cost is incurred); identifying
the targeted patient segmentation profile (based on the particular molecular mechanism); establishing new quality measures for standard of care;
setting a reasonable reimbursement price target early (determining the
molecular signature-based prescribing requirements); recruiting the segmentation population required for trial; approval in real time; and identifying indications to extend the product lifecycle. Research institutes and
other players with experience in disease biology will significantly define
and influence the choice of candidates and patients, payers, HAOs, and
even physicians will influence decisions made for the candidate prior to
full development. Regulators will become early partners and more trials
will recruit subjects and patients using social media tools. The traditional
technical infrastructure will almost certainly be impacted by cloud technologies as more basic operational activities and tasks are conducted outside of the four walls of pharma.

Like the traditional technical infrastructure, the governance model
will also change. The traditional model reflected the maxim that each network of knowledge workers along the development path assume stew-
ardship responsibility for the candidate, performing the activities to meet
milestones that will then hopefully drive the decision for approval to move
forward. As we know, time pressures (i.e., faster to market) often drive
hegemonic behavior at points along the development continuum. This
type of behavior reinforces silo mentality, making it extremely difficult to
get critical, new scientific findings or data into the governance assessment

process. Hence, critical adjustments to the plan either do not get made in a timely manner or not at all. We have seen this time and again where proof of concept is almost but never really actualized or worse when the candidate is killed far into phase III development. Even more impactful to industry and patients alike, we see this occurring after approval when critical safety risks are identified and regulators either recall or restrict prescribing (Associated Press 2012; FDA 2013b; Lee 2009).

As we have noted, the success of this new model will be limited only by how effective and efficient the codevelopers collaborate, and share risks and rewards. All stakeholders will need to maintain transparency and trust throughout the process. We recognize that not all codevelopers will share equally in the new model (i.e., financially), however their expertise will be required in other areas to make science- and data-driven decisions, balancing risk and delivering on success. All codevelopers will need a seat at the table; if not, the model will reflect what we see today: nothing more than a partner–vendor relationship model, where a partner is not really a true codeveloper.

New roles, responsibilities, and procedures required to protect privacy?

This new development model arguably still will need to address governance and how the key decision gates are realized. Governance will need to reflect the new reality that traditional activities and decisions will be made by those stakeholders outside of the company. All stakeholders will make or significantly influence decisions throughout the entire development lifecycle. To make a new model work effectively and efficiently, new roles and responsibilities must be identified for whom and when decisions are made. Roles will need to reflect codeveloper involvement and contributions based on their area of recognized scientific or business expertise.

The governance and roles and responsibilities for this new way of developing products will probably be as complex, initially, as the traditional model. Over time, however, as experience is gained through collaboration, trust, and data-driven decisions deliver results, best practices will be quickly realized and adjustments will be made to leverage from the new way of working. Whatever the model becomes, overtime, transparency, organizational buy-in, and relationship building will be critical for sustained success.

Privacy also will be an ongoing discussion for quite some time. We have seen this most recently in a whitepaper authored by several U.S. senators (Thune et al. 2013). Concerns range from "appropriateness of consent" (express authorization to disclose) for both present and future use, to security for health IT programs. However, as noted by a recent brief by the American Medical Association (n.d.) on patient confidentiality, stakeholders currently face a patchwork of federal and state laws that regulate confidentiality: the definition of breach; the consent to release; implied consent; and, public policy exceptions. When we add confidentiality to current advances in science, specifically genetic testing particularly for monogenic diseases, the landscape becomes even foggier. As genetic testing may provide us with a scientific pathway for the prevention, management, and treatment of disease, it also brings into the light additional ethical, legal, and social issues. These tests give an assessment of an individual's inherent risk for disease and disability. This predictive power makes genetic testing particularly liable for misuse. Historically, some employers and insurance companies have been known to deny individuals essential healthcare or employment based on knowledge of genetic disposition. This type of discrimination can be socially debilitating and have severe socioeconomic consequences. It is important, therefore, to ensure the confidentiality of test results, and to establish legislation permitting only selective access to this information (World Health Organization 2013). With the advent and acceptance of social media, where subscribers are volunteering to share private information, much deliberation, education, and legal reckoning will be required before patients can be assured abuse and retribution will not occur.

Concluding remarks

Within the last several years, there is an increase in prominent examples of personalized medicine (Personalized Medicine Coalition 2011). Moving toward and implementing a new product development model that can deliver on wholesale personalized medicine, however, remains an objective yet to be fully realized. The new product development model we have suggested is in juxtaposition to a current construct, and the modification and the adoption of this new way of working will require significant changes by all parties involved. The challenges are significant, and in some cases, perhaps even insurmountable. The current business model certainly can be transformed to address patent or intellectual property concerns, perhaps through patent pools or clearinghouses. Addressing the "ownership" or "contributory influence" of the stakeholders is required before any discussion of remuneration can be settled. However, information, knowledge, and data sharing/exchange and transparency are the critical threads of continuity that will need to run through the new model for it to be successful.

Although all the issues we noted will be challenging, the model's success will be dependent on a single linchpin: cultural change of the organization. This will be, by far, the greatest challenge for any stake-holder. Each entity will need to change their organizations to: embrace and be committed to transparency; openly collaborate, recognizing and respecting that others also bring critical components into the process; adopt a collective mantra for success, whereby the good of the whole supersedes the good of any single player; and that all shall share the risk. This paradigm shift, which we believe is required to deliver on personalized medicine, remains elusive at this time. In an article by Fletcher and Bourne in 2012, they describe "ten simple rules to commercialize scientific research." This is worthy of review at this time:

1. What drives science does not drive business.
2. There is no single path to commercialization.
3. You must know your rights and those of colleagues.
4. Consider the implications of going from public to private.
5. Decide how much of yourself you want to give.
6. Separate the R and the D and be realistic.
7. The market may not exist at the onset.
8. Consider the "want" versus the "need."
9. Make it comprehensible.
10. Customers are the ultimate peer review.

As noted by Fletcher and Bourne (2012), "commercializing scientific research or a breakthrough idea is really no different, in principle, from commercializing anything." Although their work focused on helping scientists get research to market, we believe their principles might be leveraged as a framework to build out the new product development model for targeted therapies and eventually, perhaps, personalized medicine. After all, the current model is not delivering today on the expectations of the new healthcare ecosystem. Temporary solutions (deals) for increasing the pipeline or revenue stream may forestall market pressure in the short term but do not appear to be viable for eventually actualizing wholesale targeted treatment therapies.

As noted throughout this chapter, the science and technology of personalized medicine is advancing at a rapid pace. The commoditizing of genome sequencing is achieved; translational research is enabling biomarker- and molecular-diagnostics-enabled therapies and biologic knowledge for normal and disease systems (positioning the approach for the clinical development realm); and data are recognized as valuable intellectual property. Information strategy is now coming into its own as industry is recognizing that their data management infrastructure is strategic—a crucial requirement—to support and enable the business in this new ecosystem.

A perfect storm is forming whereby personalized medicine may be achievable. The challenge, however, is for industry to realize that the current business model is not capable to deliver on the goal. Many in industry are or have engaged in the outsourcing of core, operational functions in order to reduce fixed costs and to satisfy market expectations—not recognizing or accounting for the extremely expensive overhead and critical, internal expertise required to affect a workable, cost-effective solution. Additionally, industry is or has acquired robust pipelines or single-candidate acquisitions, banking on near-term and future revenues. After much time and considerable spend, many robust pipeline acquisitions and or single-candidate deals (e.g., in-licensing or codevelopment deals) are found not delivering or solving only near-term revenue problems. A paradigm shift must be embraced; the model must change if personalized medicine is to be realized. The decision to move to a new development model also is not within the sole purview of the biopharmaceutical and device industry. All stakeholders operating and influencing the new healthcare ecosystem are equal players—both the model and its adoption must be communal.

References

American Medical Association. n.d. http://www.ama-assn.org/ama/pub/physician-resources/legal-topics/patient-physician-relationship-topics/patient-confidentiality.page# (accessed May 8, 2013).

Associated Press. 2012. "FDA Finds Safety Issues at Specialty Pharmacies." April 12.

Bertolucci, J. 2013. "Big Data's Human Error Problem." *Information Week*, June 10.

Byrne, C. M., Mercincavage, L. M., Pan, E. C., Vincent, A. G., Johnston, D. S., and Middleton, B. 2010. "The Value from Investments in Heath Information Technology at the U.S. Department of Veterans Health Affairs." *Health Affairs* 29:4629–638.

Denny, J. C. 2012 "Chapter 13: Mining Electronic Health Records in the Genomics Era. *PLoS Computational Biology* 8(12): e1002823. doi:10.1371/journal.pcbi.1002823.

Experts in Chronic Myeloid Leukemia. 2013. "Price of Drugs for Chronic Myeloid Leukemia (CML), Reflection of Unsustainable Cancer Drug Prices: Perspective of CML Experts." *Blood*. Prepublished online. doi:10.1182/blood-2013 03-490003.

Fleming, T., Sekeres, M., Lieberman, G., Korn, E., Wilson, W., Woodcock, J., Sridhara, R., and Perlmutter, J. 2011. "Development Paths for New Drugs with Large Treatment Effects Seen Early." Panel 4: Issue Brief, Conference on Clinical Research. November 2011. http://www.focr.org/sites/default/files/Panel4FINAL11411.pdf (accessed July 1, 2013).

Fletcher, A. C., and Bourne, P. E. 2012. "Ten Simple Rules to Commercialize Scientific Research." *PLoS Computational Biology* 8(9):e1002712. doi:10.1371/journal.pcbi.100712.

"Gleevec." 2012. *FiercePharma*. http://www.fiercepharma.com/special-reports/gleevec.

Global Edge. 2012. "Healthcare: Introduction." http://globaledge.msu.edu/industries/healthcare (accessed May 8, 2013).

Gold, E. R., Piper, T., Morin, J.-F., Durell, L. K., Carbone, J., and Henry, E. 2007. "Preliminary Legal Review of Proposed Medicines Patent Pool." http://www.theinnovationpartnership.org/data/documents/00000003-1.pdf (accessed May 8, 2013).

Green, C. S., and Troyanskaya, O. G. 2012. "Chapter 2: Data-Driven View of Disease Biology." *PLoS Computational Biology* 8(12). doi: e10.1371/journal.pcbi.1002816.

"HIE Market Report: Analysis & Trends of the Health Information Exchange Market." 2011. Chilmark Research. www.healthit.gov (accessed May 10, 2013).

Hsiao, C.-J., and Hing, E. 2012. "Use and Characteristics of Electronic Health Record Systems Among Office-based Physician Practices: United States, 2001–2012." NCHS Data Brief, No. 111.

Huttenhower, C., Hibbs, M., Myers, C., and Troyanskaya, O. G. 2006. "A Scalable Method for Integration and Functional Analysis of Multiple Microarray Datasets." *Bioinformatics* 22:2890–2897.

"Informed Consent form from the Harvard Medical School's Personal Genome Project." 2011. http://www.personalgenomes.org/consent/PGP_Consent_Approved03242009.pdf (accessed May 8, 2013).

Jamoom, E., Beatty, P., Bercovitz, A., Woodwell, D., Palso, K., and Rechtsteiner, E. 2012. "Physician Adoption of Electronic Health Record Systems: United States, 2011." NCHS Data Brief No. 98. July.

"Kaiser Permanente, UCSF Scientists Complete NIH-Funded Genomics Project Involving 100,000 People." 2011.

Lee, M. 2009. "Safety, Efficacy and Ethical Issues with Weight Loss Medications." *Clinical Pharmacology* 2(2):111–113.

McCaughan, M. 2012. "RPM Report—"Breakthrough Therapy": New Pathway in FDASIA May Point the Way to Future Reforms." Friends of Cancer Research. http://www.focr.org/news/rpm-report-breakthrough-therapy-new-pathway-fdasia-may-point-way-future-reforms.

Metcalfe, T. 2009. "Regulation of Biomarkers Used in Personalised Medication." http://www.eurobio-event.com/DocBD/speaker/pdf/75.pdf.

Moore, G. 2013. "Here's Why AstraZeneca CEO Pascal Soriot is Betting Huge on a Tiny Cambridge Biotech." *Boston Business Journal*, March 21. http://www.bizjournals.com/boston/blog/mass_roundup/2013/03/astrazeneca-moderna-investment.html?ana=e_bost_cap&s=newsletter&ed=2013-03-21&u=11020419364f105fd795e3b350c9d4 (accessed May 8, 2013).

Novartis. n.d. http://dominoext.novartis.com/nc/ncprre01.nsf/0/d2c9464baf-c5b55bc1256ff500609fee (accessed May 8, 2013).

Parsons, H. 2007. "Personalized Medicine Inevitable, Says Lilly's Lechleiter." *PharmaManufacturing.com*, December 18. http://www.pharmamanufacturing.com/articles/2007/177.html.

Personalized Medicine Coalition. 2011. "The Case for Personalized Medicine 2006 and 2012." http://www.ageofpersonalizedmedicine.org/objects/pdfs/PM_by_the_Numbers.pdf (accessed May 8, 2013).

Rothman, S., Ravels, V., Friedman, A., and Rothman, D. 2011. "Health Advocacy Organizations and the Pharmaceutical Industry: An Analysis of Disclosure Practices" *American Journal of Public Health* 101(4):e1-38. doi:10.2105/AJPH.2010.300027.

Swan, M. 2012. "Crowdsourced Health Research Studies: An Important Emerging Complement to Clinical Trials in the Public Health Research Ecosystem." *Journal of Medical Internet Research* 14(2):e46.

Thune, J., Alexander, L., Roberts, P., Burr, R., Coburn, T., and Enzi, M. 2013. "Reboot: Re-Examining the Strategies Needed to Successfully Adopt Health IT." Whitepaper, United States Senate, April 16.

U.S. Department of Health and Human Services (HHS). 2012. "HHS Retrospective Review Update." http://www.hhs.gov/open/execorders/13563/retrospectivereviewchart2012-09.pdf.

U.S. Food and Drug Administration (FDA). 2013a. "Center for Devices and Radiological Health: 2013 Strategic Priorities." http://www.fda.gov/AboutFDA/CentersOffices/OfficeofMedicalProductsandTobacco/CDRH/CDRHVisionandMission/ucm330378.htm (accessed May 8, 2013).

U.S. Food and Drug Administration (FDA). 2013b. "Human Drug Product Recalls Pending Classification." (Recalls from 2004-2012.) http://www.fda.gov/Safety/Recalls/EnforcementReports/ucm310739.htm.

Wetterstrand, K. A. n.d. "DNA Sequencing Costs: Data from the NHGRI Genome Sequencing Program (GSP)." National Human Genome Research Institute, www.genome.gov/sequencingcosts.

World Health Organization. 2013. "Genetic Testing." http://www.who.int/genomics/elsi/gentesting/en/# (accessed May 8, 2013).

Index